Gender, Science and Technology: Perspectives from Africa

Edited by

Catherine Wawasi Kitetu

CODESRIA Gender Series 6

CODESRIA

Council for the Development of Social Science Research in Africa

Typeset by Hadijatou Sy

Printed by Imprimerie Graphiplus, Dakar, Senegal

ISBN: 978-2-86978-221-1

The Council for the Development of Social Science Research in Africa (CODESRIA) is an independent organisation whose principal objectives are facilitating research, promoting research-based publishing and creating multiple forums geared towards the exchange of views and information among African researchers. It challenges the fragmentation of research through the creation of thematic research networks that cut across linguistic and regional boundaries.

CODESRIA publishes a quarterly journal, *Africa Development,* the longest standing Africa-based social science journal; *Afrika Zamani,* a journal of history; the *African Sociological Review; African Journal of International Affairs* (AJIA); *Africa Review of Books; Identity, Culture and Politics: An Afro-Asian Dialogue* and the *Journal of Higher Education in Africa.* It copublishes the *Africa Media Review.* Research results and other activities of the institution are disseminated through 'Working Papers', 'Monograph Series', 'CODESRIA Book Series', and the *CODESRIA Bulletin.*

CODESRIA would like to express its gratitude to the Swedish International Development Cooperation Agency (SIDA/SAREC), the International Development Research Centre (IDRC), Ford Foundation, MacArthur Foundation, Carnegie Corporation, NORAD, the Danish Agency for International Development (DANIDA), the French Ministry of Cooperation, the United Nations Development Programme (UNDP), the Netherlands Ministry of Foreign Affairs, Rockefeller Foundation, FINIDA, CIDA, IIEP/ADEA, OECD, OXFAM America, UNICEF and the Government of Senegal for supporting its research, training and publication programmes.

Contents

PART III: Science and Technology:
The Case of One Woman, Many Women

List of Contributors

Anthonia Achike, Department of Agricultural Economics, University of Nigeria, Nigeria.

Catherine Wawasi Kitetu, Department of Languages and Linguistics, Egerton University, Kenya.

Damian Opata, Department of English, University of Nigeria, Nigeria.

Elisabeth Sherif, Centre d'Etude d'Afrique Noire, Institut d'Etudes Politique, Dedeaux, France.

Fibian Kavulani Lukalo, Institute for Human Resource Development, Moi University, Kenya.

Ghislaine Agonhessou Yaya, Department of the NGO FMN, Calavi, Benin.

John Wilson Forje, Department of Political Science, University of Yaounde II, Cameroon.

Jackline Kerubo Moriasi, Department of Agricultural Economics and Business, Egerton University, Kenya.

Lydia Ayako Mareri, Department of Languages and Linguistics, Egerton University, Kenya.

Mweru Mwingi, Department of Education, Rhodes University, South Africa.

Kenneth Nyangena, Department of Development Support, University of Orange Free State, South Africa.

Olubukola Olakunbi Ojo, Department of Educational Foundations and Counseling, Obafemi Awolowo University, Nigeria.

Samson Moenga Omwoyo, Department of History, Archaeology and Political Studies, Kenyatta University, Kenya.

Acknowledgements

First and foremost, I give many thanks to CODESRIA, which invited and financed the one-month peer reviewing of the proposals presented at the Gender Institute 2003. CODESRIA also funded all the research that the laureates went on to do, and finally arranged the publica-tion of these papers. Thanks to the CODESRIA secretariat in Dakar, who worked tirelessly with us while in session with much cheerfulness and good humour. Thanks are also due to the resource persons and all the laureates of the Gender Institute 2003, for their patience and very active participation during the workshops and, later, while working on their papers which form the contents of this book. Many heartfelt thanks are also due to my co-director, Josephine Beoku-Betts, who painstakingly reviewed the first drafts of all the papers in this collection. All efforts have been made to polish the thoughts presented here. However, the positions adopted are those of the writers of the chapters and not necessarily those of the co-directors of the Gender Institute 2003 or CODESRIA.

Introduction

This collection is a contribution to current debates on gender, science and technology. Feminist debates and research about science can generally be put into two slots: the women question in science (that is, women's participation in sciences) and the science question in feminism (that is, the construction of feminine knowledge). The chapters in this collection deal mainly with the first—women's participation in science.

In recent years, science and technology has been widely advocated as the indisputable foundation for political and economic power in the modern world. However, science is still marked by various layers and dimensions of deep-seated gender inequality that work to the disadvantage of women. Despite the fact that a lot of awareness has been created and that gender issues are now more readily acknowledged by developers in Africa, participation in science and technology continues to remain a hurdle as far as girls and women are concerned. Half the African population is lagging behind in science and technology. This is not lost on developers, but there is presently a dearth of research to show what actually goes on in the region. UNESCO provided seminal information through its regional office for Education in Africa, based in Dakar (BREDA 1999) in a project titled 'Technical, Scientific and Vocational Training for Young Girls', which identified factors determining how positively or negatively girls were being guided into scientific and technical streams. Measures adopted by member states to improve girls' access to these streams were also identified. However, UNESCO's initiatives have not received adequate follow up. CODESRIA's Gender Institute of 2003 was therefore a step in the right direction towards filling this gap. The chapters in this book, which are the outcome of that institute, are a welcome contribution.

The findings of the UNESCO-funded project showed that gender inequalities in sciences are not inevitable. In fact, where public authorities, teacher associations and officials decided to take action, positive results regarding girls' involvement were obtained, such as in 'science clinics' in Ghana and Nigeria, Olympiads and *blocs scientifiques et technologiques* in Senegal. The present collection sheds light on situations on the continent and factors that hinder such positive actions.

A common theme runs through the chapters: the exclusion of girls and women from science and technology in Africa is about feminine identities, ideologies of domesticity and gender stereotypes. There are no clear policies on gender and science in most countries. Education practices follow societal beliefs and help define feminine identities, which then reproduce the ideology of domesticity among girls and encourage a rejection of science and technology. Later in life, these ideologies translate into stereotypes that result in women being kept away from scientific positions and being denied access to technical jobs.

Issues of gender and science and technology are not witnessed in African countries alone. Few countries in the world have managed as yet to deal with these issues adequately. However, when Africa is involved, as is the case with everything else, these issues become enormous. A critical examination of feminist studies shows the voice of the African woman sadly lacking not only in science and technology but in everything else. Most feminist studies have been blind to the contradictions and complexities that inform girls' experiences. The task therefore is to reclaim the African woman from, not only exclusion from sciences, but also religion and other social areas. Exclusion is even affirmed in African mythology, where proverbs and all wise sayings stress gender differences and divisions.

Some Pertinent Questions about Science and Technology

What are science and technology? Science is concerned with how and why things happen, while technology deals with making things happen. Science explains the reasons behind phenomena, while technology is the use of tools, machines, materials, techniques and sources of power to make work easier and more productive. However, dominant science, as it is practiced today, adheres to Northern and Western conceptualisations, values and language use. As such, science is historically and culturally (read Western) located. It is not value-free.

Science should be informed—must be informed—by the daily experiences of people in the world. Yet, in most parts of Africa, the experiences of girls and women are excluded. This collection examines this exclusion and generally points at education policies and practices and cultural perceptions as the main hindrances to girls and women's participation in science and technology. However, there are many questions that still need answers: how do Third World scientists position themselves in the face of Western dominant science? How do African scientists interpret scientific meanings and technologies practised by their people? Are there other ways of 'knowing' (real African ways of knowing) that are lacking in science as it is taught today? Why African women? What distinctive approaches do women bring to science?

There are arguments that the knowledge of an African woman comes from her interaction with nature through, for example, subsistence agriculture and seed germination, as exemplified by initiatives such as the Green Belt movement in Kenya championed by Professor Wangari Maathai (Nobel Peace Laureate 2004). However, this knowledge is not documented, and neither are women's contributions incorporated into development initiatives. The writers of this volume, mainly drawn from the social sciences because of these subjects' strength in gender studies, make their contribution by showing how exclusion takes place and how much still remains to be done to overcome this exclusion. Questions related to science knowledge—and whether there is a body of knowledge which can be called African science and technology—await elucidation, together with those posed above.

The chapters are grouped into three parts. Part I, 'Science and Technology in Society: Discourse, Perspectives, Practices and Policy', consists of three chapters. The first chapter, by Catherine Kitetu, is a kind of ground-clearing text for the other chapters. It starts by assessing the importance and place of science and technology today as key for socio-economic development in the world. It goes on to show the packaging and discourse of science that leads to the exclusion of Africans generally, and African women in particular. The discourse of science, that is, the language use, beliefs and ways of structuring knowledge that are upheld in society and in institutions where science is taught, are critiqued. The process of how discourse of science in schools, the institution through which science is perpetuated, positions women and girls to reduce their access to science and technology, is also outlined. Using an example from Kenya, it is shown how the wider society forms the base of much of these school-gendered practices. Other issues, such as foreign knowledge, worldview and languages used to transmit science for both African males and females, plus the mode of application of science and technology research in Africa, are raised for further debate. The chapter specifically calls for more qualitative analyses in Africa.

The second chapter, by John Forje, is on 'National Science Policy'. It is mainly a position paper that focuses on ways in which science and technology policies on the one hand, and research and development policies on the other, can be articulated and aggregated. Forje addresses these notions from a platform of thought provoking questions, many of which should goad scholars into lively debate. He acknowledges that little has been done to incorporate either science and technology or research and development into national development plans, and argues that African governments must stimulate and sponsor the promotion of science and technology through a comprehensive and concerted science and technology policy under the auspices of a national development strategy. Africa faces problems of pressing food shortages, illiteracy, energy insecurity, water supply, health care, shelter, communication and environmental problems, but it can contribute to the global scene by integrating science as we know it today with Africa's own inherent cultural heritage and indigenous knowledge systems.

Damian Opata, in his chapter on 'Binary Synthesis, Epistemic Naturalism, and Subjectivities', examines what he calls unwarranted antinomies, binary oppositions and adversarial knowledge politics that he sees as bedevilling gender scholarship in the West and which is being transported to Africa. He brings in the perspective of 'binary synthesis' (or complementary dualities) that, among the Igbo of Southeastern Nigeria, depicts an ontological order based on the mutual co-existence of things that are as they are. He argues that this perspective is necessary for gender studies in Africa and draws inspiration from the Igbo ontological nature of existence in which nature is seen as a model for technological development, and in which ordered subjectivities are framed on the notions of autonomy and individuation rather than on gender. Opata's ideas invite debate. For example, one could consider the fact that, while Igbo culture recognises duality and male and female have each their spaces, yet the female cannot 'fly' as she wishes due to cultural constraints, many of which are not for her good.

Part II deals with science and technology in educational contexts. Science and technology are transmitted through education, and the exclusion of women from the sciences starts in the early phases of their education and through the kind of motivation they receive. This section starts with a chapter by Elisabeth Sherif in which she identifies the political and institutional causes of under-representation of girls. She analyses the actions and strategies adopted by the government of Niger in its regulation of the education system. Like many other African governments, it has been highly dependent on external financing in recent years. Sherif analyses, not only the government's political choices, but also its relations with international actors, such as the World Bank, in the context of the application of structural adjustment plans. The thrust of her argument is that, despite their apparent gender neutrality, educational policies have an impact on girls' access to scientific and technical education. She reiterates that the under-representation of women in science courses and programmes is an obstacle to development because it guarantees the reproduction of the cycle of gender-linked inequality and helps maintain women in a situation of socio-economic dependence. Sherif poses a question at the end, which should carry the debate on gender, science and technology further: Would the disastrous side of science have developed if women had participated in the design and implementation of science and technical projects?

The next chapter, by Ghislaine Agouhessou Yaya, is a critical analysis of how different policies for the education of girls in Benin have been carried out. She shows the socio-anthropological factors that determine and influence the selection of scientific subjects at school and university, and argues that Benin's education policies have helped steer female students away from scientific subjects at secondary and university levels. Yaya identifies the family, teachers and the labour market as key determinants of subject choice among students. As a way forward, she proposes re-training teachers, sensitising parents and re-examining the procedures of the council charged with placements of students.

Olubukola Olakunbi Ojo then shows how school counselors in Nigeria can and should be engaged productively in streamlining students into science and arts classes as part of the placement service that they provide in school settings. Ojo sees the role of counselors as indispensable in relation to girls' decision to enter science-related programmes. Counselors influence girls' perceptions of science and technology consciously, subconsciously and unconsciously. Guidance and counseling should be an indispensable arm of education, but sadly, in the survey she carried out, few counselors saw the need of encouraging girls to aim at science careers.

Kenneth Nyangena's chapter examines the views, on science and technology, of students in a high school in Kenya. He shows students' own role in the persistent under-representation of the female population in education generally, and science specifically. He speculates that, if school science were made more 'girl-friendly', that is, centred on girls' interests and ways of working, this would be one step towards creating a feminine science. And if considerable numbers of girls (and boys) emerged from school, having learned science in this way, perhaps science as an institution would begin to change.

Mweru Mwingi's case study of girls' participation in science and technology education in Murang'a district in Kenya is a further examination of science in school. Her argument is that, while there are numerous studies that link negative attitudes to low female participation and poor achievement in science and technology subjects, few studies examine how accessible school subjects are within the school curriculum. For example, subjects in the secondary school curriculum in Kenya are said to be equally accessible to all; but in practice, access to science and technology subjects depends on the kind of school one attends. School curriculum implementation guidelines and socio-structural factors permeate subject choice.

Lydia Ayako Mareri looks at the handling of science in schools' extra-curricular activities. Recommendations have been made in policy documents in Kenya to enable curriculum materials to be re-designed to ensure that they are relevant to girls and boys and that teachers are sensitised to treat girls and boys equally. However, Mareri argues, this would not be possible without packaging this also in the extra-curricular programme. Focusing on out-of-class activities within the school setting, she discusses the roles that girls and boys are assigned in the performing arts (plays, dramatised dance, dramatised verse and oral narratives). She uses the performing arts items usually presented during the annual Kenya National Drama Festivals and finds that schools' out-of-class socialisation activities exclude girls from performances which contain science notions.

Does theatre matter? Does participation of girls and boys in specific roles matter? The question is whether girls can be influenced to do science through such activities. While the jury is still out, it can be argued that extra-cururicular activities can provide opportunities for both girls and boys to familiarise themselves with science notions. More importantly, theatre can help to change the masculine image of science and also the stereotyping of girls as being unsuited for science careers.

Fibian Kavulani Lukalo provides another discussion on curriculum subject choices. In her chapter, she shows how the need for increased science and technology in Kenya led to the introduction of computer learning in schools in 1994, but how the categorisation of computer as a science subject led to poor participation by girls. Computer studies were mapped against the societal backdrop of gender biases which, when put together with geographical disparities in school type, availability of resources and infrastructure (electricity, water, laboratories, land and space) and teaching personnel, led to many problems from the start. It came as no surprise that only two girls were registered for the first national examinations in computer studies in secondary schools in 1998! Kavulani concludes that educational policy, with respect to computer studies, reflects gender disparities and gender role expectations in the wider society, which adversely affects the participation of women and girls.

Part III: 'Science and Technology: The Case of One Woman, Many Women', adopts a historical perspective. This section shows that, while it is true that women have been excluded form science and technology, and their accumulated experiences and knowledge made invisible in the human scientific patrimony, yet women have never ceased devising clever and ingenious ways to enable them to master nature, though most of the time away from the limelight. The chapters in this section give us a glimpse of this history. Moriasi examines the life of Sudan's first woman surgeon. She showcases this woman as a rare role model for African girls and argues that oral history can be a basic tool in our efforts to incorporate the previously overlooked lives, activities and feelings of women. When women speak for themselves, they reveal hidden realities. New experiences and new perspectives emerge that challenge the 'truths' of official accounts and cast doubt upon established theories. Interviews with women can explore private realms such as reproduction, child rearing and sexuality to tell us what women actually did, can do or should have done. This chapter urges oral historians to develop techniques that will encourage women to say the unsaid. It is from this perspective that she traces the life and times of Nahid Toubia who, as the first woman surgeon in Sudan, conquered great odds to become what she is today.

Samson Omwoyo then assesses the impact of coffee production on Abagusii women of Western Kenya. His chapter blends two important themes: the changing role of women in agricultural production and the impact of new agricultural technologies. Women were disadvantaged in the pre-colonial period, and the oppressive and exploitative relationship between men and women in pre-colonial Gusiiland was amplified by technological innovations accompanying the introduction of coffee in the area. The lucrative cash crop was often the preserve of male farmers, while female farmers were relegated to subsistence crop production. Research into high-yielding varieties and the use of pesticides and fertilisers benefited male farmers, thus entrenching them in 'modern' agriculture, with female labour. Omoyo analyses the response of women in coping with this marginalisation and their methods of survival.

The last chapter in Part III, by Anthonia Achike, looks at gender-based associations (GBAs) and their role in enhancing the participation of female farmers in science and technology projects in Anambra State, Nigeria. Drawing her data from a survey study of five GBAs, she shows how they have responded to the plight of disadvantaged female farmers by having more female-targeted science and technology innovations in their portfolios. Unfortunately, however, female GBAs attract less external financial assistance from governments and international organisations when compared to their male counterparts. Hence, female GBAs have less technical and financial potential to assist female farmers despite their stronger grassroots support.

This is in line with the age-old thinking where, almost by instinct, women are considered to be less capable of creating science and technology.

The research presented in this book is characterised by the researchers' own chosen topics based on their geographical areas of operation. Although studies from the Northern and the Horn of Africa are missing (there were no participants), the chapters provide a window, albeit a limited one, onto the current state of female participation in science and technology in Africa, along with analysis of the historical backgrounds and discussion of what needs to be done in the future. Clearly, more research urgently needs to be done with more groups throughout Africa. It is our hope that these studies will inspire more qualitative work within the continent in relation to gender science and technology.

Catherine Wawasi Kitetu

Part I

Science and Technology in Society: Discourse, Perspectives, Practices and Policy

1

Discourse and Practice of Science: Implications for Women in Africa

Catherine Wawasi Kitetu

Introduction

The focus of this chapter is the role of science and technology in socio-economic development in the world today. It examines the 'packaging' of science that leads to the exclusion of Africans generally, and of African women in particular. The differences between men and women in terms of how they access and use science and technology is now well documented (Harding and Hintikka 1983). Most studies show that women are excluded not only from science classrooms but also from scientific and technological fields and professions. Feminist research in science studies has been two-pronged—with one branch handling 'the women question in the sciences' (i.e., women's participation in sciences) and the other 'the science question in feminism' (i.e., the construction of 'feminine knowledge'). There have been lots of debates and research on women in the sciences but few empirical studies on construction of feminine knowledge. Not enough answers have been provided for such questions as whether gender-differentiated approaches to understanding of science can be observed or whether male and female students, in tackling investigative problems, display different 'images' (understandings) of scientific knowledge.

The literature on girls and science is vast, ranging from theoretical speculations and interpretations (Kelly 1981, 1987; Walkerdine 1989) to empirical research (Crossman 1987; Spear 1987; Kelly 1985; Whyte 1986). These studies are, of course, part of an even wider body of literature concerned with gender in education.

Recently, women's participation in sciences has been tied to identities. Gendered identity refers to the sense of self, including the conscious and unconscious thoughts and emotions of the individual as a gendered being. Studies have shown that girls may position themselves as not being interested in sciences and technologies because doing science is 'for men' and therefore not the 'correct' identity for a girl. Some of these constructions are very subtle, as I will show shortly. Such feminine-gendered identities are said to reproduce the ideology of domesticity among girls.

Gender is a socially constructed attribute of an individual based on sex. It has a pervasive influence on us whether we like it or not. There is always gender differentiation in any context; one is seen as either male or female and treated as such. Such differentiation does not necessarily translate into discrimination or bias. There is reason to believe, for example, that the rights of some African women have actually been safeguarded through gender differentiation. However, with modernisation, the notion of gender has become synonymous with sexism and its negative connotations of discrimination, usually of the female gender. Nevertheless, gender is a complex category that is continuously variable; it is not fixed but changing.

This instability of gender enables us to explore ways of changing the participation of women in science and technology. As Cameron (1997) aptly says:

> Gender is regulated and policed by rather rigid social norms, but this does not mean ... [men and women are reduced to automata, programmed by early socialisation to repeat forever the appropriate gendered behaviour ... [T]hey are conscious agents who may engage in acts of aggression, subversion and resistance. As active producers rather than passive producers of gendered behaviour, men and women may use their awareness of the gendered meanings that attach to particular ways of speaking and acting to produce a variety of effects.

Science and Technology in Society

What is science? This question can be answered in various ways. When posed during the Gender Institute for Laureates in Dakar in 2003, the question caused a heated debate. Questions were even asked as to whether Africans *have* any science. Can different cultures have different sciences? Is it possible to talk of multiple sciences when referring to human knowledge? If so, whose science were we really discussing in this conference? This paper does not address such fundamental questions but instead pursues the generally agreed-on notion that, when one talks about science and technology today, one thinks of the 'high-tech' science practiced from and by the Western/Northern nations. In this chapter, I will be referring to this idea of science without going into the contentious debates mentioned above.

This Western science is first and foremost a discourse of 'technology', defined as 'a way of thinking/speaking and doing things' (Halliday and Martin 1993). Science in this sense has driven industrialisation and the development of new technologies and has consequently acquired special significance in society. Today's debates over post-modernity, technology and the globalisation of politics, economics and culture are posited upon this science. As Luke (1992) argues:

> What counts as 'science' in the period since World War II has been focal in the development of Western Nation States, to the point where historical winners and losers in economic, strategic and geopolitical realism are assessed in terms of technology and scientific prowess.

Luke traces the dominance of science in society to the late nineteenth and early twentieth centuries when scientists (i.e. laboratory scientists), government institutions and corporations began to work together in Europe and America and the results of laboratory research began to be adapted for practical purposes by government and business. It is from this time that applied approaches to physics, mathematics, statistics, electronics and computing, communications and engineering emerged. Our modern technocratic society has its basis in this symbiotic relationship between governments, corporations and research science, particularly after World War II. The process has shaped the character of everyday life, from mass consumer culture to research and academic institutions. Science and technology, in the worldwide context, have become the dominant mode for interpreting human existence: everything 'from the discourse of technocracy and bureaucracy to the television magazine and the blurb on the back of the cereal packet is in some way affected by the mode of meanings that are science' (Luke 1992).

Science and technology have also determined the international distribution of wealth and power. Today, dependence on corporate science and technological expansion is a key means for the expansion of state power and legitimacy. Scientific nations translate into rich economies, while nations that lag behind in science and technology translate into poor economies. The yoking together of state-funded military industrial complexes with corporate capital and academic scientific research has been witnessed in modern wars, such as those in the Middle East, but has also been witnessed in the global economy, in HIV/AIDS research and in reproductive technologies. While it should not be forgotten that some of this research has been at the centre of heated ethical debates and controversies, for example, the science of cloning, in general, what has come to count as science in technocratic culture is science that is applied, corporate and profitable.

Africa does not seem to fit very comfortably in this overall picture. It is a consumer continent, buying its technology from the West while providing the West with raw materials to produce this technology. African scientists and government institutions do not appear to apply or make profitable the outcomes of their laboratory research. Yet the African leaders who led the fight against colonialism realised that the European monopoly of science enabled and perpetuated colonialism. At independence, science was promoted as a central factor for development and is still equated with power and cultural worth. Thus, science subjects continue to be very important in education throughout Africa.

The question is whether science education in African nations has been successful in developing scientists, particularly when we consider gender. How much of the science learned is actually put to use? Below, I turn to science education, using Kenya as an example to consider issues of science and gender in schools where, supposedly, the new crop of scientists are being raised. First, I examine male dominance in science and in education generally.

Male Dominance in Science, Technology and Education

Science is transmitted formally through schooling, but there has been a concern that, in this transmission, girls and women are not adequately involved. Science is overwhelmingly male-dominated in terms of who is or can be a scientist. This is true not only in Africa but throughout the world, and this gendering of science has therefore been a global concern for some time. The problem has been approached from many different angles. One has been a broadly psychological angle that tries to understand why girls avoid sciences and looks for answers in girls' attitudes and personality traits. In this approach, girls are seen as the source of the problem. Their own psychological characteristics are seen as contributing to their failure to perform well in science. A second approach is from a sociological and structural angle in which science is seen as a socially constructed process, a set of practices produced in schools in accordance with societal norms. Since society is an institution that assigns men and women different roles, with different obligations and expectations, this division is also mapped onto school subjects.

Male Image of Science

The sciences, especially the physical sciences, are overwhelmingly male-dominated in terms of who teaches it, who is recognised as a scientist and also the way science is packaged (Kelly 1985). Most textbooks thus portray scientists as men (Obura 1991). The 'masculinity' of science inheres in the form and content of scientific knowledge and in the underlying belief that this knowledge can only be successfully pursued by men and boys. Easlea (1986) has shown that scientists are expected to be aggressive, individualistic, self-confident and competitive. Various cultures associate men with competence in the design and control of apparatus and machines, while women are associated with competence in activities demanding care and nurturing. Thus, it is no wonder that science is associated with the image of masculine ability.

The masculine image of science is further seen in the way in which the scientific method of inquiry expects scientists to be guided by logic and facts alone. Scientists are expected to pursue a ruthless analysis of reality exemplified by the suppression of all emotion or wishful thinking in order to arrive at the goal of 'truth' about nature. However, such methods are now being contested as not wholly desirable, as they lack 'feminine' intuition and feeling, yet intuition can lead to scientific discovery. Even more striking in modern science is the extent to which male scientists portray nature, the supposed object of scientific inquiry, as metaphorically 'female'. Merchant (quoted in Easlea 1989) suggests that one of the most powerful images in science is the 'identification of nature with a female, especially a female harbouring secrets' (1989). A scientist who can 'woo' secrets out of nature is acclaimed as a genius. This imagery can be related to the larger problem of patriarchal discourses, thinking and knowledge that exclude females (Easlea 1986, Walkerdine 1989, Harding and Hintikka 1983, Rose 1994).

Nilan (1995) brings an interesting angle to this imagery of science. She talks of the gendering of school subjects in which gendered identities are connected with an orientation towards clusters of subjects perceived as either 'hard' and 'masculine' (physics, maths, economics) or 'soft' and 'feminine' (humanities, arts). An important question researchers have asked is what kind of influence this masculine image has on students, particularly girls. Many researchers have suggested that this image is the root cause of the under-representation of women in the scientific world. In fact, students are said to perceive science as masculine from a very early age (Kelly 1981, 1987).

Studies in science education have been carried out by physical scientists as well as by researchers in many different disciplines—sociolinguistics, sociology, history, feminist studies and philosophy of science—and this is now a broad and increasingly diverse field of study. In the social sciences, the focus has been on the institution of science, the practices of scientists and the nature of scientific knowledge, i.e., the question of what constitutes scientific knowledge (Rose 1994). In linguistics, the problem is seen as the language of science and how scientific knowledge is constructed in language (Myers 1990; Halliday and Martin 1993).

Discourse of Science

The discourse (language use) of science subjects needs examination. Millar (1989), Driver (1989) and Solomon (1989) have shown how classroom science is a social activity involving teachers and students and have stressed that language plays a crucial role in forming and consolidating ideas. In classroom talk, teachers need to pay attention in order to achieve a balanced participation of girls, particularly when interactions are initiated by pupils. More importantly, however, science language is a specialised discourse used in specific social situations. Halliday and Martin (1993) have traced the evolution of science discourse over the years as a 'discourse technology'—a linguistic semiotic practice developed in order to do specialised kinds of theoretical and practical work in social institutions. As a discourse technology, science has a special significance. Science discourse and its dominant practices dictate a lot of human activities. Halliday and Martin (1993) argue that this discourse has become the language of literacy for the elite and notes that the discourse of science has evolved over the years in contrast to the discourses of folk wisdom. Othiambo (1972) argues that science discourse ignores folk wisdom, which Africans are closer to.

Science discourse has its own methods of inquiry, language and imagery. Some of the prototypical features of scientific discourse, according to Halliday and Martin (1993), include 'objectification' (i.e., actions, events and qualities presented as if they are objects), a high concentration of content words (i.e., complex technical taxonomies within a single clause), passive forms, nominalisation and ambiguities (i.e., a string of nouns leaving inexplicit the semantic relations among them). Such forms of language can be unfriendly to most people. In short, science language is different from ordinary language. Science language is used to express relationships of classification, taxonomy and logical connections. As a genre, it is not a narrative like literature or history. It is special because of its content, its written and spoken genres, its activity structures and styles. For example, there is much use of the impersonal passive voice. People tend to disappear as actors and agents. Colloquial language, personifications, figurative language, irony, humour and exaggeration are avoided. Fiction and fantasy give way to talk of 'facts' (Lemke 1990). As such, scientific literacy and the mastery of what is called scientific knowledge is the exclusive domain of a few. It is elitist. Scientific practices end up excluding women, ethnic minorities within the Western world and whole groups of people in the southern hemisphere, mainly because of this specialised science discourse.

Specifically, what puts students, particularly girls, off in sciences? Most studies have shown that learning science means learning the conventional forms of organising scientific reasoning, talking and writing. Doing science also involves conducting investigations, using apparatus, observations and measurements. In a science class, activities such as demonstration and exhibition of physical and chemical processes are prevalent. In her study of science practical work done by students, French (1989) observed that, from the procedures used to the language and the final conclusions, science classroom practice adheres to the meanings and values of science as an institution. Science is treated as objective truth.

Lemke (1990), as part of extensive studies of science classroom practice, has looked into what science values mean to learners. In a detailed study of the science classroom, he observed that science was consistently presented as 'objective' knowledge. In the process, as he argues, 'more than just science was being communicated in these classes.... [A] set of attitudes towards and beliefs about science, education, students' abilities, and society itself were being taught as well'. Lemke shows that science is difficult for students not only because of its technical aspects but also because it creates what he calls a 'mystique of science' that sets up a pervasive and false opposition between the world of science—objective, authoritative, impersonal— and the ordinary world of human uncertainties, judgments, values and interests. Thus, he argues:

> ... it is not surprising that those who succeed in science tend to be like those who define the 'appropriate' way to talk science: male rather than female, white rather than black, middle- and upper-middle class native English-speakers, standard dialect speakers committed to the values of North European middle-class culture (emotional control, orderliness, rationalism, achievement, punctuality, social hierarchy, etc.).

As a result, Lemke observes:

> It is only people whose backgrounds have led them to speak more like science books do, learn in a particular style, and a particular pace, already have an interest in a certain way of looking at the world and in certain topics and problems that will have a chance of doing well in sciences.

Lemke's argument brings to the fore some questions that are especially pertinent to Africa. What language should be used to teach science to African children? Should science in African schools continue to follow that worldview and topics? If scientific language and thinking can be hard to grasp even for certain males, how much worse must it be for African females? Sadly, few studies have been done in the science classroom in Africa to examine these issues, and most have involved survey/statistical methods rather than in-depth qualitative classroom studies. Below, I turn to one example of the latter.

Gender and Science in a Kenyan Classroom

This Kenyan study (Kitetu 1998) opens a small window into the interactions of gender and science education in Africa inside the classroom and points to how we can begin to rethink science and technology research and application. The study involved research conducted over three years and produced diverse findings regarding the handling of science as a subject, classroom management, use of language and construction of gendered identities. In this chapter, I can only provide a glimpse into these findings.

The study was prompted by the problems facing girls and sciences in Kenya, voiced by different interested groups. A few studies had already been done, but none had addressed science classroom discourse and gender. Two key questions guiding the study, were: 'why girls were not involved in the sciences and why the few who were involved performed so dismally'. Focusing on language and other practices of the physics classroom, the study set out to examine the construction of gender in the physics classroom. Physics was chosen because it was perceived as being the most gendered of all the science subjects in the curriculum. Using an ethnographic approach, classroom interaction and laboratory activities were observed and audio-recorded. Participants were then interviewed to get their perception of the activities they were involved in. This was a descriptive study. Statistical methods have not been very fruitful in showing the social relations that obtain in science education as far as gender is concerned, but the ethnographic approach in this study helped unearth subtle gendered behaviour.

The study showed that science teaching and learning adhered to the discourse of global science in terms of the genre of science, i.e., in relation to learning the conventional formats for organising scientific reasoning, talking and writing. Doing science in these classrooms involved conducting investigations, using apparatus and making observations and measurements. Activities such as demonstrating and exhibiting physical and chemical processes were prevalent. This is an area where more research and debate should be done in order to discuss the science appropriate for the African child. This would mean looking at the language in use, the content and the application expected from this learning. The study showed that concepts were often too foreign, especially when explained in a foreign language and, moreover, in a scientific form of that language. The tools and materials were also often unfamiliar.

The students, however, were expected to fit into this discourse. A telling remark from one female student interviewee confirmed this. On failing a class test, she was asked where the problem was and replied:

> ...I think I explained it here .... Maybe I explained it the way I explain in other subjects *where you use your mind* .... Maybe I was expected to explain in physicist terms. (Emphasis mine)

This student had the feeling that science was different from other subjects. In science, one was required to remember things as they were taught, not in one's on way or by using one's own mind! Science needs to be demystified and taught as a normal part of human experience. This was not happening here.

Gendered identities were constructed in both verbal and non-verbal activities. Teachers addressed girls and boys differently. Boys were talked to in harsher ways (i.e., confrontational tones, e.g., 'why did you?...', said loudly) in contrast to the softer tones used on girls. Also, in the laboratory, the boys and the teachers did the 'hard' experiments (i.e., measuring, cutting, constructing) for the girls, thus gendering the girls as different or weak. The girls also fitted into this positioning without question. They did the traditionally feminine, 'soft' activities such as reading instructions, writing and washing up after experiments. One teacher explained: 'You are not supposed to be harsh on girls but we have to be friendly so they do not feel that they are being punished by doing physics'. The girls interpreted this treatment as being 'favoured', yet such positioning has serious implications for girls' advancement in science; if they are not 'pushed' like the boys, they are likely to fall behind.

This gendered construction of the physics classroom was clearly based on the wider society's perceptions of feminine and masculine gender roles and identities. Kenyans generally are very careful in the upbringing of girls. A 'properly' brought-up girl (in terms of feminine attributes) is 'protected' and cannot go places on her own, even if it is safe, since this does not project the right image. Science, and education generally, do not fit into this cultural picture. This thinking is changing, but very slowly.

This Kenyan study urgently requires not only duplication with similar groups throughout the region but also in terms of the world view obtaining within Africa in relation to science, technology and language. Research should also focus on the gendered discourses (beliefs). For example, it would be good to know how cultural identities of femininity (e.g., 'a good girl stays at home or gets home early') are in opposition to science and school identities (e.g,. 'good students work long at a science project in the school laboratory'). Research should also look into culturally determined gender interpersonal relations (e.g., 'members of opposite sexes don't sitwork together') in opposition to school interpersonal relations such as working in mixed-sex groups or teachers of different sexes giving individual student tuition. Research should also handle the cultural gender role of girls as 'mother's helpers', and see how this is in opposition to school roles, such as having enough time to do homework.

Conclusion

This paper has examined how gender is implicated in science, showing how research all over the world has shown women to be uninvolved or excluded from science. I have also shown how science, as taught in schools (the main way of perpetuating science), excludes certain groups of people, mainly groups in the southern hemisphere and women. The Kenyan study shows in a small way how gender identities are constructed while doing science in school; girls remain girls and boys remain boys and are treated as such, not as learners or upcoming scientists.

In sum, I should note that issues surrounding science and technology are complex. There are issues of discourse and practice (discourse as ways of thinking, ideologies, values and meanings that surround science and also discourse as images and language). There are also concerns regarding the way science is taught in schools, such as the foreign languages used and the worldview advocated, not to forget the whole issue of globalisation, which complicates matters further. The Kenyan study provides a small window on how African teachers and students are grappling with the sciences (the language, the experiments, the topics and the gendered roles) and shows how teachers and students draw much of their understanding from the society in terms of interpreting not only the science concepts but also the worldview, gender roles and identities. It is along all these facets of gender science and technology that recommendations are made for African feminists to address analytic skills.

References

Cameron, D., 1997, 'Performing Gender Identities: Young Men's Talk and the Construction of Heterosexual Masculinity', in S. Johnson and U. Meinhof, eds., *Language and Masculinity*, Oxford: Blackwell, pp. 50-60.

Crossman, M., 1987, 'Teachers' Interaction with Girls and Boys in Science Lessons', in A. Kelly, ed., *Science for Girls?* Manchester: Manchester University Press.

Driver, R., 1989, 'The Construction of Scientific Knowledge in School Classrooms', in R. Millar, ed., *Doing Science: Images in Science Education,* London: Falmer Press.

Easlea, B., 1986, 'The Masculine Image of Science with Special Reference to Physics: How Much Does Gender Really Matter?', in J. Harding, ed., *Perspectives on Gender and Science*, London: Falmer Press, pp. 132-158.

Halliday, M. A. K. and Martin, J. R., 1993, *Writing Science: Literacy and Discursive Power,* London: Falmer Press.

Harding, S. and Hintikka, M. B., 1983, *Discovering Reality*, Boston: Ridel.

Kelly, A., ed., 1981, *The Missing Half: Girls and Science Education*, Manchester: Manchester University Press.

Kelly, A., ed., 1985, *Girls into Science and Technology (GIST): Final Report,* Dept. of Sociology, University of Manchester.

Kelly, A., ed., 1987, *Science for Girls?* Manchester: Manchester University Press.

Kitetu, C., 1998, 'An Examination of Physics Classroom Discourse Practices and the Construction of Gendered Identities in a Kenyan Secondary School', unpublished Ph.D. thesis, Lancaster University.

Lemke, J. L., 1990, *Talking Science: Language, Learning and Values,* Norwood, NJ: Ablex Publishing.

Luke, A., 1992, 'Series Editor's Introduction', in M. A. K. Halliday and J. R. Martin, 1993, *Writing Science: Literacy and Discursive Power,* London: Falmer Press.

Millar, R., ed., 1989, *Doing Science: Images in Science Education,* London: Falmer Press.

Myers, G., 1990, *Writing Biology: Texts in the Social Construction of Scientific Knowledge,* Madison: University of Wisconsin Press.

Nilan, P.,1995, 'Negotiating Gendered Identity in Classroom Disputes and Collaboration', in *Discourse and Society*, Vol. 6, no. 1, pp 27-47.

Obura, A., 1991, *Changing Images: Portrayal of Girls and Women in Kenyan Textbooks,* Nairobi: Acts Press.

Othiambo, T.R., 1972, 'Understanding of Science: The Impact of the African View of Nature', in P. G. S. Gilbert and M. N. Lovegrove, eds., *Science Education in Africa,* Nairobi: Heinemann.

Rose, H., 1994, *Towards a Feminist Transformation of the Sciences,* Cambridge: Polity Press.

Solomon, J., 1989, 'The Social Construction of School Science', in R. Millar, ed., *Doing Science: Images in Science Education,* London: Falmer Press.

Spear, M. G., 1987, 'Teachers Views about the Importance of Science for Boys and Girls', in A. Kelly, ed., 1987, *Science for Girls?* Manchester: Manchester University Press.

Walkerdine, V., 1989, *Counting Girls Out: Girls and Mathematics,* London: Virago.

Whyte, J., 1986, *Girls into Science and Technology: The Story of a Project,* London: Routledge and Kegan Paul.

2

National Policy on Science and Technology: An Integral Component of Development Strategy for African Countries

John W. Forje

Introduction

Africa cannot develop without constructively addressing issues that incorporate the proper development, application and utilisation of science and technology for sustainable development. If the quality of livelihood of the vast majority of the population is to be improved, drastic measures must be taken now to ensure the cultivation, growth, nurturing and utilisation of science and technology as integral components of national development strategies. Up till now, little has been done to incorporate science and technology (S&T) or research and development (R&D) as integral parts of development plans, yet no country has developed by bypassing science and technology. The way forward is for Africa to begin deploying its indigenous knowledge systems together with modern technology for the socio-economic transformation of the continent.

This paper looks at worrying trends in Africa, in relation to science and technology, but also outlines a way in which national policy can be formulated as an integral part of a development strategy that also incorporates gender. Several questions need to be posed regarding national policies for science and technology in Africa. What capacity do African countries have to build national systems for S&T and R&D policy in universities and research institutes? To what extent is the private sector willing to follow government S&T and R&D policies? How can the private sector contribute to S&T and R&D development? What role do political factors play in shaping state capacity? Do political parties have S&T and R&D policy agendas? What is the role of the international community? How should women be incorporated into S&T and R&D policy formulation, development and utilisation? What is meant by S&T and R&D policy? In what ways can scholars share experiences and learn from each other to jointly contribute to knowledge creation, the necessary platform for empowering women and putting them on the development agenda during an era guided by the unstoppable forces of globalisation, information and communication technologies (ICTs) and genetic engineering? Do the people of Africa *want* to develop? What kind of development is envisioned for them? For whom, by whom and for what purpose is development in Africa to be pursued? How do we want to develop? By what means, when and how should the development come?

These questions are posed mainly to provoke discussion. They cannot be answered here, but some definitions of key terminologies are put forward, together with a survey of scientific and technological capacity in Africa, gender issues in science and technology, key theoretical perspectives and the formulation and implementation of national policy. The chapter ends with suggestions for a way forward.

Definitions of Key Terminologies

First of all, what are science and technology policies? Science and technology policies are forms of state intervention intended to promote the development and dissemination of knowledge and the practical application of research results to the production of goods and services. Science and technology policies contribute to building national innovation systems by introducing new technologies as well as improving on existing indigenous knowledge systems. Policy guidelines should act as a compass to pilot the nation out of underdevelopment, poverty, corruption, exploitation and marginalisation. Policy choices depend on government's capacity for obtaining and using relevant information, and particularly its capacity for taking into account a complex set of social interests while at the same time keeping a degree of independence from social and political pressures. The underlying challenge is how to translate national initiatives and international cooperation in science and technology into concrete activities, programmes and processes. Opportunities abound as Africa responds to its current state of scientific and technological underdevelopment, and a secure technological future can be built by taking advantage of information communication technologies (ICTs) and biotechnology/genetic engineering. The continent can also draw invaluable lessons from the economic history of the Asian 'tigers', many of which, as recently as the 1960s, were far more underdeveloped in terms of GNP per capita than some African countries at the time. A key factor in the successful industrial and economic transformation of these countries was that they were able to formulate and implement policies to harness science and technology for development. Their experience shows that policy implementation depends on administrative competencies as well as on broad political support and consensus at various levels. It requires government intervention to mobilise appropriate information and to develop the consensus necessary for policy implementation. In short, the success of any given policy depends on partnership, participation, responsibility and benefit-sharing between the state, civil society, the productive sector and the international community, as I argue below. African countries need technological dynamism to address underdevelopment predicaments in three key areas: food production, plant breeding (including fertilizer and pesticide production) and health care systems. To face these challenges, policy guidelines, along with the support of high-level political institutions, in other words, the political will, are crucial.

Lack of Scientific and Technological Capacity

Many African countries do not have comprehensive, coherent science and technology policies. Elsewhere in the world, policies are clear, and industrial productivity is driven by science and technological inventions based on such policies. In most African countries, however, there are weak links between policy, science and industrial activities:

> Local industries generally purchase technology and related know-how from abroad rather than connect to local scientific thrusts. Towards the lower end of the development spectrum there is an almost total disjuncture between science and industry (Adam 2001).

Knowledge too is lacking. The *World Development Report* (1998) on the global explosion of knowledge identifies both threats and opportunities in this regard:

> If knowledge gaps widen, the world will be split further, not just by disparities in capital and other resources, but also by the disparity in knowledge. If we can narrow those knowledge gaps, it may be possible to improve incomes and living standards at a much faster pace than previously imagined (World Development Report 1998).

Gender in Science and Technology

Within the African continent, policy and politics constantly disadvantage women in relation to science and technology. For sustainable, science-driven development to take place, the relationship must move from the minus to the plus side of the development continuum. However, the relationship between scientific and technological change on the one hand and socio-economic development on the other are interrelated and complex. While it is true, as Wad (1982) notes, that there are many differences of opinion about what works and what does not work in terms of gender, science and technology policy, it also remains clear that policy plays a key role in relation to the what, when, how, why and for whom science and technology is done.

Theoretical Perspectives

There is no clear agreement on the most workable theoretical framework for science and technology policy. One possible framework is system theory, which, according to Easton (1953), is a response by the political system on input and output processes and the decisions brought to bear on it. Fundamental within this conception is the identification of institutions and activities that function to transform demands (inputs) into authoritative decisions (outputs) necessitating the support of society. Elite theory, on the other hand, sees S&T and R&D policy as reflecting the interests and values of elites, not the demands of the people or the needy. From this perspective, public funds tend to be used not for the interests of the masses but for those of a particular class. Another theoretical framework, group theory, focuses on establishing the 'rules of the game' for the selection of issues, making different groups accountable for their policy preferences, enacting policy options by compromising and balancing various group interests. Policy here depends on consensus, which in turn depends on the relative influence of the different groups. Rational choice theory is a framework drawn from economics that depends on the concept of rational choice, while development theory (also called modernisation theory) offers a critique of traditional models of decision-making constrained by time, cost and intelligence. Finally, language theory emphasises the importance of language for enhancing, promoting and disseminating innovations in appropriate technology to the masses. It stresses that how we communicate with one another is vital in the development process. An appreciation of all these theories should inform policy formulation.

Formulation of Science and Technology Policy

At independence, many developing countries embarked on five-year national development plans, which were intended to provide a basis for the full incorporation of S&T and R&D as integral parts of the national development process. However, over the years, these five-year plans fell by the wayside of the (under-)development process. In their place, ad hoc policy measures and strategies became the order of the day, reducing S&T and R&D to an insignificant role. To date, numerous national and international factors have impacted on state capacity in the formulation and implementation of S&T and R&D policy. Science policy-related activities in African countries have in general evolved almost from scratch in many sectors to fit economic and political goals. The role of government in science policy has varied from country to country and over time.

As a result the practice of S&T and R&D policy in many African countries can be described in terms of:

- gross absence of political will and focus on development
- confusing policies on science and technology
- unrealistic goals
- inability to implement agreed-on policy recommendations
- inadequate financial inputs for S&T and R&D activities.

Where different forms of policy instruments exist, no explicit technology policy can be said to have been effectuated due to non-articulation and non-implementation of policies in respect of the following interrelated factors:

- poor understanding of the relationship between technology and economic, industrial and social development
- unrealistic understanding of the nature of dynamic industries and their technological requirements
- dependency and structural imbalances in the economy
- absence of political will and of indigenous strategic programmes to address underdevelopment.

In contrast, the newly industrialised economies of Asia and Latin America have adopted explicit technology policies directed at addressing the critical areas of:

- managing technology transfer
- managing technology change
- developing technological capacity.

Moreover, in Africa, the knowledge needed to transform existing natural resources into finished consumer products is lacking. It is vital for African countries to formulate policies capable of nurturing these critical areas instead of remaining raw material-exporting countries, Policy should be aimed at developing the capacity to assimilate, adapt and disseminate technologies as well as to innovate and be creative in building up an indigenous scientific and technological base. Human knowledge is extremely important. This explains why resource-poor countries like Denmark, Luxembourg and many others can still dictate to resource-rich African nations.

Therefore, the content of science policy must first be concerned with supporting both research and the utilisation of research findings. In formulating a science policy, at least three themes need to be considered:

- Scientific and technological innovation and the relations between government and industry, so that the necessary conducive environment for development can be put in place
- Issues of disparity between national research and development efforts, so that research results can be translated into consumer goods
- Creating the necessary framework for international scientific networking, so that the quality of domestic research is improved.

Secondly, science and technology policies and policy instruments must be specific in order to address and focus on:

- human resource development and institutional capacity building, which should involve illiteracy eradication, education campaigns and high-quality educational services and infrastructures
- social consensus through participation and political pluralism, i.e., democratic governance and good management incorporating the values of civic responsibility, discipline, stability and ethical values
- incorporation of socio-cultural factors in technological development through the proper use of resource potentials, cultural diversity and indigenous

knowledge systems as input assets, thus promoting and sustaining an indigenous science and technology culture

- development of strong incentives for the commercialisation of R&D results.

Thirdly, for society to move forward by means of technological discoveries and innovations, there is need to marry the four broad functions, the scientific, the professional, the administrative and the political, of government and public affairs. At one end of the spectrum, pure science is concerned with knowledge and truth. At the other end of the spectrum, politics is concerned with power and action. But none ever exists in its pure form.

Institutional malfunction, inadequate theoretical knowledge, poor understanding of technological problems and decision processes, inadequate research resources and inappropriate research methods must end. The academic community must get out of its ivory tower, stop prostituting itself for political posts and, instead, show its professional ethics by addressing the needs of the community. Individual citizens must be prepared to participate in decision-making about technological change, technology policy and good governance.

Elements of a Framework of Action

The most important element for a framework of action is commitment on the part of the government, civil society and the productive sector to make S&T and R&D integral parts of the overall national development strategy. There has to be a committed political will from the state and other actors. The orientation of national development planning should take account of long-term technological developments and requirements. The major requirements of S&T and R&D development include:

- development of national technological capabilities for technology policy, technology appraisal, diffusion and application, institution building, human resource development and effective exchange of information
- survey and identification of natural resources
- prompt and easy access to relevant technological information
- creation of a demand for local technological capabilities.

An important step in this direction is to put in place interactive structures. The first task is infrastructure development. Building the technological infrastructure capacity base is very important both within the productive sector (private sector) and in government (public sector), universities and research institutes. Universities, research institutes and basic research programmes, along with administrative mechanisms for their funding, coordination and control, have to be established. Multidisciplinary technological research institutes and services related to natural resources, such as meteorology services, oceanographic and geological surveys and mapping/remote-sensing agencies, are also required along with service institutes for testing, standards, quality control and troubleshooting. Finally, the local engineering and consulting industries must be promoted.

These activities require a sustained political commitment at the highest possible level of government, coupled with constant vigilance to ensure that the quality of performance meets international standards and is not undermined by the problems that typically beset government bureaucracy. Also, funding of research should be given top priority. As of now, this constitutes a weak link in the advancement of science and technology in many African countries. University-private sector linkages should be encouraged, with the major industries sponsoring research-related activities in their areas of interest.

Popularisation of S&T and R&D Policies

To popularise science and technology policies, all aspects of communication and dissemination mechanisms have to be developed and deployed in order to:

- find a common language to overcome cross-sector communication barriers and to generate increased understanding of scientific, socio-economic and management issues
- provide science and scaling tools to allow for policy-making aimed at management processes and/or procedures on different geographical and societal levels
- make science responsive to national issues through better linking with socio-economic, cultural, industrial, political and development factors
- bring science and technology nearer to the people and develop indigenous knowledge as inputs in the development process
- use information communication technologies (ICTs) adequately.

The challenge is for the mass media to articulate means and ways of getting the right message to the people on the role and impact of S&T and R&D in the socio-economic transformation of society and as a means of quantitatively and qualitatively improving their living standards.

Political parties, civil society and the general public, especially young people and women, should be encouraged to develop an interest in science and technology disciplines such as mathematics, physics, chemistry, biology and others. Groups, parties and institutions should also have their own policies on science and technology based on the national one. Specifically, this calls for policy measures that:

- promote national and cross-border cooperation and connectivity by utilising both indigenous knowledge systems and the modern knowledge currently available in existing centres of excellence on the continent
- develop and adapt information collection and analysis capacity to support productive activities
- generate a critical mass of technology expertise in targeted areas that offer high growth potential, especially in biotechnology and geo-science
- make use of the expertise of Africans in the diaspora to spur the process of change on the African continent.

Attention should also be paid to the Lagos Plan of Action (LPA), Africa's most articulated strategic policy framework, which failed to take off due the economic crises of the 1980s, and which today has been transformed into the New Partnership for Africa's Development (NEPAD). The LPA calls on African countries individually and collectively to give top priority to science and technology for sustainable development and to move from policy to action-oriented development activities with a human face.

Ways Forward

Science and technology policy in itself is not a panacea for resolving all the problems of the African continent, but it has the potential to transform their economies and achieve sustainable development through the quality and effectiveness of each country's policies and institutions that support science, technology and innovation. It is a means to an end and depends on the vision of leaders, the intensity and quality of interactions among the various stakeholders, the dedication of R&D bodies and policy-making departments and the dynamism of private economic enterprises that determine whether a country or particular region can harness science and technology for development. African countries should also learn from the successes of other countries, particularly in relation to:

- strong democratic government and committed leadership as essential conditions for economic, technological, scientific and industrial-take-off
- state participation in scientific and technological development, as in South Korea, Singapore, Brazil, India, Taiwan and others
- expansion of the education sector while tailoring education to address the specific needs of the society with an emphasis on technical education
- private-sector investment in human-resource capacity-building
- promotion of a culture of hard work, innovation, creativity, responsibility and accountability.

Additionally, just as an isolated human being is unable to share his or her ideas with anyone else or benefit from the ideas of others, a country or continent that tries to stand alone cannot make significant advances in addressing the needs of its people, nor can it take advantage of the rapid advancements in the domain of science and technology in other parts of the world. Africa should not be isolated in an age of globalisation and information communication technology. The move by Nigeria to launch a satellite with the aid of the Russian space station is most encouraging. On the whole, this may be the century to make, or continue to mar, the future of the people of Africa.

References

Adam, R., 2001, 'Choosing Good Science in Developing Countries', paper presented at the IST Roundtable on Africa, Science and Technology in the Age of Globalisation, Nairobi, August 2, 2001.

Wad, A., ed., 1982, *Science Technology and Development*, London: Westview Press.
World Development Report, 1998, *Knowledge for Development,* Washington, DC: World Bank.

3

Binary Synthesis, Epistemic Naturalism and Subjectivities: Perspectives for Understanding Gender, Science and Technology in Africa

Damian U. Opata

Introduction

CODESRIA's 2003 Gender Institute on Gender, Science and Technology was an exceedingly important one. Science and technology have, for more than three hundred years, been the most important factors shaping the modern world. In the process, they have revolutionised gender relations in both the home and the workplace. As Alvin Toffler (1999:9) notes, 'the most important economic development of our lifetime has been the rise of a new system for creating wealth, based no longer on muscle but on mind.' A new knowledge system come into being, resulting from the study and practice of science and technology. Before Toffler, Shirley Burggraf (1997) had written:

> Two hundred years is just a blink in evolutionary time, but within two centuries, we have developed from a frontier economy in which women were dependent on men for economic and physical survival (hunting, tilling the land, fighting) to an industrial economy with care taking, clerical, retail, and processing jobs at which women could support themselves at a low level but were still dependent on man for earning better wages and for fighting wars, to a post industrial economy based on knowledge, information, and service skills at which women seem to be as naturally adept as men.

In spite of this major and irreversible development in the social division of labour along gender lines, the momentous revulsion over gender inequity felt by Western feminists has continued to be steadily and increasingly stimulated since at least the mid-twentieth century. Consequently, gender and feminist studies, which easily slide into one another, have been appositional, adversarial, robustly political and therefore power-driven, especially in the West.

The intention of the CODESRIA 2003 Gender Institute to interrogate the role of women in science and technology in Africa can best be understood against this background. This is why this paper already situates itself in a doubleness of relation to the other: first, to the role of women in science and technology as, for instance, against that of men; and second, to the role of women in S&T in Africa as against that of women in S&T in the West. Both issues connote some essentialisation and homogenisation of women in S&T at the same time as they highlight differences on a global scale, where they exist, and subvert the idea of the fundamental sameness of the human species, that is, the elements that make both men and women human. Admittedly, gender studies, as already suggested, announce themselves as partial, but this partiality need not be seen in adversarial and binary oppositions, as is the case in West.

For this reason, the perspective I want to bring to bear in this paper is that of binary synthesis, of complementary dualities, which, among the Igbo of Southeastern Nigerian, for example, depict an ontological order that displays the mutual co-existence of things that are as they are. This perspective is necessary for gender studies in Africa so that scholarship in the area, while not legitimating any inequitable gender relations, does not spawn the unwarranted antinomies, binary oppositions and adversarial politics that have bedeviled gender scholarship in the west. Consequently, the paper draws inspiration from the Igbo ontological view of existence in which all beings need one another in a harmonious relationship, in which nature is seen as a model for technological development and in which ordered subjectivities are framed on the notions of autonomy and individuation rather than on gender. These notions are respectively spelt out in the three parts of this paper.

Part 1, Binary Synthesis, Gender, and Existence among the Igbo, deals with the Igbo notion that *ife di abuo abuo; nkaa kwulu, nkaa akwudebe ya* ('Things exist in twos; when one thing stands, the other stands beside it'). Part 2, Epistemic Naturalism in Science and Technology among the Traditional Igbo, examines the concept expressed in the proverb *okpu uzu lee egbe anya na-odu* ('Let a would-be smith look at the tail of the kite'). Part 3, Subjectivities and Gender Relations Among the Traditional Igbo, focuses on the Igbo saying *one an nke ya, onye na nke ya* ('Each person to his or her abilities'). The term 'epistemic naturalism' deserves some explanatory remarks. In using this term, I draw inspiration from three philosophical doctrines: foundationalism, naturalism and naturalised epistemology. I use the term as both a model of learning and as the name of a concept that makes knowledge and its transformation possible by making nature the norm and source of amassing knowledge that can be used to design and manufacture instruments that enhance human productivity. The methodology adopted in this paper is largely 'archival' and interpretive. By archival in this context, I mean going back to folklore, particularly as represented by proverbs, in order to use this knowledge as a site for the retrieval of cultural memory. The proverbs that are 'mined' in the process serve as quasi-theoretical postulates that share in the contestation to stabilise the world by offering certain principled statements about the way the world is. As forms that posit and at the same time share in the contestation for meanings, they make themselves subject to and available for interpretation. In using the archival and interpretive method, I follow a path trodden by many social scientists and literary scholars. In the specific area of gender and feminist research, I follow the examples of Mary Daly, Clarisa Pinkola Estes, M. Esther Harding and Charlene Spretnak, to name just a few who have drawn extensively from myths, legends and folktales in their works.

Let me begin with an act of transgression constructed to typify the Igbo understanding of binary synthesis. Among the Igbo of Southeastern Nigeria, the tradition of kolanut breaking, along with its accompanying ritual, has long become anecdotal. In the process, such widely accepted statements that are often preparatory, even apologetic, have evolved. Two outstanding examples of such statements are: 'he who brings kola brings life' and 'kolanut does not understand English'. The latter is uttered by way of apology and deference to the use of the Igbo language when the kolanut is being broken in a multi-lingual setting (Opata 1988). The act of transgression that I want to commit lies in my use of English to perform this ritual. I will be offending the ancestors of Senegal, whose language is not Igbo, but as the act is not committed within Igbo land, I feel confident that when I get back home, I will settle the matter with my ancestors.

The Act of Transgression

Our people say that he who brings kola brings life.

Oh God, come and take Kola!

The land of Senegal, come and take kola

The land of Nigeria, come and take kola

The land of Africa, come and take kola

All land come and take kola; for all land is one

Human beings responsible for assigning some parts to some people.

Let the Sky come and take kola, and let the Earth come and take kola

Let there be life abundant for what is up, and let there be life abundant for what is below

Let man live, let woman live

Let there be life for fish, let there be life for water

Let the kite perch and let the eagle perch, the one that says no to the other may its wings break

The person who has come to our place, let him not bring us bad luck; when he departs, may he not grow hunchback

Let our life be blessed, and let the life of those that come after us be more blessed

Let what we will eat come, let what will kill us depart

We have come to Dakar in peace, we shall go in peace.

Let CODESRIA live, let the participants of the 2003 Gender Institute live

In this Institute, we shall encounter no evil spirits; we shall encounter no evil people

If what will destroy us is behind, let us be in front; if it is in front, let us be behind

What we meet in Dakar that is good, we shall take home; what is bad in

Dakar we shall leave behind us

All this we ask in God's name

This act of transgression is not really new. It is something that must be done in the service of scholarship. Chike Aniakor (1979) did a similar thing, thus:

Ancestors take kola

We are only children who wash ourselves around the belly

Give us life

Give us children

Give us wealth

With which to raise our children

Let the kite perch

Let the eagle perch

Any that says no

Let the wings be clipped

If one says yes, the Gods echo their support.

The principle of complementary duality is evident in the text, but Aniakor's analysis goes to point out that if one has life, one has need for children, if the kite perches, the eagle shall also have a perching space, etc. One thing stands in relation to another. This is what foregrounds binary synthesis among the Igbo.

Binary Synthesis in the Igbo Worldview

The acts of transgression just depicted represents the fundamental Igbo orientation to life, capturing the idea of binary synthesis or complementary dualities so evident in Igbo life and thought. The fact that this ritual takes place, in a more or less elaborate form depending on the context, each time kola is broken among the Igbo serves not only to propagate the tradition but also to reinforce its importance, especially among the younger generation. This fundamental perspective is best captured in the Igbo proverb *ife di abuo abuo, nkaa kwulu, nkaa akwudebe ya*, meaning 'things exist in twos; when one thing stands another stands besides it'. This proverb is further strengthened by another proverb to the effect that 'what is one is a finished thing' (*ife di naa gwuru agwu*). Any kolanut that has no lobes is thrown away as an evil object. Any hen that hatches only one chick is killed along with its young. The simple reason is that such occurrences break the ontological perception of the Igbo about the nature of existence.

In a seminal work, *After God Is Dibia: Igbo Cosmology, Healing, Divination and Sacred Science in Nigeria* (1997), J. A. Umeh explains that the number one is the unique attribute of God, whereas the number two represents *okwu na abo chukwu kwulu*, i.e., 'the two words known by God'. According to Umeh, the Igbo give a more detailed explanation of the two spoken words of God in the following manner:

Nwoke na nwanyi (men and woman/male and female)

Onwu na ndu (death and life)

Oku na mmili (fire and water/energy and liquid)

Uchichi na eshishe/efife (day and night).

Kpakpando di n enu na aja di n'ana/kpakpando na aja di n' ana

(stars above and sand on the ground)

Enu na ana (heaven and earth/up and down/sky and ground/above and below).

Oso na ije (running and walking).

Ula na imu anya (sleeping and being awake)

Obele na nnukwu/ukwu na nta (tall and short).

Ocha na oji (white and black).

Ji na ede (yam and coco-yam)

Be mmuo na be mmadu (spirit home and human world).

Nnu na ose (salt and pepper).

And so on to infinity, pairings continuing without end.

It is no surprise, therefore, that this complementary duality cannot but frame the way the Igbo think about all facets of existence. Umeh, a Professor of Estate Management but also chief priest to the Idemili shrine in Nobi, Southeastern Nigeria, a diviner and herbalist, adds an explanatory note to these endless pairings:

[T]he number 2 is encapsulated in the Igbo mystical idiom of 'the two words of God', namely *eeye/Aa* (yes) and *ee - ehimba* (no). It is only human beings that add a third word namely: *Ee - ee - eh* (yes and no or yes-no), which introduces obscurity, falsehood, deception, fraud and other evils' (1997:34).

While not wishing to challenge Umeh's mystical insight, for which I have no competence, I would add that human beings' addition of a third word is an indication of human limitations, of the inability to make definitive statements that can only emanate from God. This would also seem to account for the conceptual emphasis on complementary dualities rather than on binary oppositions, because limited human knowledge is best understood within the notion of complementarities rather than oppositions. One can conclude by noting that Umezinwa and Animalu (1988), along with others, have also observed the pervading sense of binary synthesis or complementary dualities among the Igbo.

A very important principle, which results from this overarching sense of complementary duality among the Igbo, is that of connectedness. The principle of connectedness among things (existents) ensures that some type of synergy resulting from the natural release and correspondent absorption of energy by existents takes place. Trouble starts when the kite perches but refuses the eagle a perching space, when the kite, in other words, occupies more space than its size and needs require, thereby encroaching on the perching space of the eagle. It is in order to avoid such trouble that the principle of accommodation becomes not only necessary but also a guiding principle of living among the Igbo.

The Place of Gender Within the Igbo Worldview

One of the creation accounts in Genesis talks of how God created male and female. Among the Igbo, things that exist in the world are posited as *oke na nne*, that is, male and female. Any creature that is neither male nor female is abominable, although male creatures can be castrated and lose their maleness. Even plants exist as male or female. The famous Igbo kola also has male and female lobes. Musical instruments have male and female counterparts. Natural phenomena like rain and sunlight are also described in male and female terms. It is interesting to note, however, that *oke and nne are* not only denotative but also connotative. Thus, *oke* also means 'great', while *nne* can also mean 'caring'. Let us look at the following examples:

oke mmadu	great human being
oke nwanyi	great woman
oke mmili	heavy rain/great rain
oke anwu	severe drought
oke obodo	great town
iji oke ugwo	being greatly indebted,

On the other hand, *nne nwanyi* can mean 'mother' or 'caring woman', a female who encapsulates the essence of womanhood through motherhood and caring. We can also have *nne mmadu* meaning, again, a woman who has mothered a child. The use of *nne* as mother is evident in such Igbo expression and names as *nnenna* (mother of [one's] father), *nne mmuo* (a woman old enough to be accorded some privileges reserved for men who have been initiated into masquerade cults) and *nne-ukwu* (grandmother or godmother).

The word *oke* is further used to depict male virility and the subjugation of the female, despite the evident binary complementarity, which does not, in any case, mean equality. This virility is shown in the connotative use of the word *oke*. In traditional societies, and still today, when a female animal is on heat, the Igbo describe the condition as *icho oke*, that is, 'searching for male'. The animal is then taken to a place where there is a male counterpart. Here, the female is tethered to a rope, while the male is let loose. When the female animal has been served, it is taken home. From this condition of the she-animal on heat, the Igbo derive the term *iche mmadu n'oke*, implying that when a human being behaves as if he or she is on heat, the person is subdued, put in his or her proper place or made to behave in a normal way. The metaphor is very important, because an animal on heat behaves irrationally, impulsively and in an uncontrollable manner. Thus, *iche mmadu n'oke* is a metaphor for control and subjugation. This connotative use of *oke* does not apply to the word *nne*.

The word *oke* is also used to depict the physical superiority (in terms of strength) of the male over the female. Thus, the Igbo proverb *mbe n'abo nuo ogu, a mara nke bu oke* (when two tortoises fight, the male one will be known) means that when two tortoises fight the male one will overcome and illustrates this Igbo understanding of the greater physical strength of males. Opata (2003) has done a more extensive study of this proverb. In addition, the derogatory Igbo word for a bachelor is *oke-okporo (okokporo)*, which literally translates as 'empty male'. This is not just to suggest that a man ought to marry, but that a man who fails to marry has not actualised his maleness. This may further suggest that the essence of being *oke* (male) is to have a woman under one's care.

But if the word *oke* is denotatively and connotatively used in Igbo language to depict maleness as greatness, virility and even superiority, pronominalisation in Igbo language is gender-sensitive. Igbo pronouns are not genderised. Whereas most European languages have feminine and masculine forms of the pronoun, Igbo language has just one common pronoun. Thus, there is no parallel to the English usage of 'man', 'mankind' and so on to generically refer to both male and female persons. This is a particularly important point, because such a linguistic structure embeds the subordination of woman, but among the Igbo, this is not the case.

One other important feature among the Igbo when it comes to naming is that a woman never loses her maiden name when she gets married. The contemporary situation in which a woman changes her name after getting married to a man is a cultural importation. This situation makes a woman lose her ancestral identity and take on the identity of the husband. The phenomenon of hyphenated names can only be a manifestation of the conflict attendant upon altered identities after marriage.

Binary Synthesis, Gender and Notions of Equality

Equality, it must be stated, is a nebulous concept. The ideal it is supposed to capture is highly problematic. What is evident in nature everywhere is an overwhelming manifestation of inequality in both the size of things and the opportunities and capacities they have to survive. The political and constitutional notion of equality as developed over time in Western scholarship is totally inadequate for a meaningful understanding of gender relations among the traditional Igbo.

The Igbo say *ishi na ishi ha bu n'onu*, literally meaning 'one head is equal to another only in the mouth'. In other words, the equality of human beings is only a verbal affirmation, not a real condition. If all human beings cannot be equal, it will be a false statement to affirm in Igbo that *nwoke na nwanyi ha* (male and female persons are equal) because, as the Igbo also say, *onwere nwanyi ka nwoke*, i.e., there are instances of women who are greater than men.

Consequently, I want to abandon the use of the concept of equality and settle for such other things as rights and opportunities. These rights and opportunities no doubt exist in a socially structured society with dominant organisational hierarchies. Within what has been explained as binary synthesis in the Igbo worldview, there is a need to find out how the rights and opportunities of individuals define their self-actualisation. To start with, among the traditional Igbo, there are things allowed to people as rights. These rights take two major forms: rights due a person on account of age or status, and rights due to person's sex. Some of the rights are not seen in legal terms but in a normative way, in that the society feels that such things ought to be. Let me give an example. In my village, Umoda, Lejja, Nsukka local government area of Enugu state, when-ever there is a social gathering and there is palm wine, it is the right of younger persons to be served palm wine before older ones. In the same vein, if there is meat in the food being served, it is the right of an older person to be served meat before a younger person. It is through this complementarity of rights across age ranges that social harmony is achieved. It is also to be noted that these rights go with responsibilities, so that, whereas the older persons are concerned with contemplation and legislation, the younger ones are more engaged in action and implementation of social and customary policies. What is important here is the idea of complementarity.

There are also rights and responsibilities spread along gender lines. Women cannot preside over ancestral deities; they cannot wield the traditional mace, they cannot inherit ancestral land and they do not perform kolanut rituals, at least in terms of petitioning ancestors, etc. The reasons for this are to be found in the patrilineal and patrilocal social structure predominant among the Igbo. Within this arrangement, it would appear illogical for a woman, for instance, to preside over an ancestral deity, since the woman's ancestors belong to another town or community. It is not that women are incompetent to perform such functions—after all, there are women priestesses among the Igbo—only that the deities they serve are not ancestral. On the other hand, women have rights to farmland in their place of marriage and rights to certain parts of slaughtered animals. Just like men, they have specific rights to specific economic trees, etc. For instance, a child growing teeth in certain Igbo communities is entitled to one palm tree, but the right to harvest fruit from the tree devolves on the grandmother. This means that a woman has rights even from the reproductive labour of her daughters. But what is important here again is not who has more rights, but the manner in which the rights are exercised. The rights are exercised to complement one another, perhaps in an asymmetrical relationship to the responsibilities.

The opportunities for self-development of both males and females are again largely defined by patrilineality and patrilocality, as well as by perceptions of gender roles as dictated by compelling social realities. This is evident in early childhood, when children are socialised into doing different things according to their sex, not according to their abilities, which have not yet been determined. To the traditional Igbo, it made no sense, for example, for a father to take his daughter with him to the farm and ask the son to remain behind to help the mother with domestic work. It made more sense for a father to teach a son the 'masculine' things and for a mother to teach the daughter 'feminine' things. It was largely a matter of learning by gender association. The father prepared the son to be a father, and the mother prepared the daughter to be a mother. The important thing was that fatherhood and motherhood were seen as complementary, not oppositional.

Epistemic Naturalism and the Practice of Technology Among the Traditional Igbo

The Igbo proverb *okpu uzu lee egbe anya na—odu* (let a would-be smith take a look at the tail of the kite) has a variant that goes like this:

> *uzu amaro akpu ogene ya nee egbe anya n'odu* (the blacksmith who does not know how to fashion the twin iron gong should observe the tail of a kite)

Umeh (1997), who documents this variant, goes on to make the following explanation of the proverb:

> Whatever your profession or calling may be, if your receptive and perceptive faculties are open and alert, you will observe that nature is the supreme adept and the ultimate. Nature is the best scientist, the best architect, and the best estate surveyor and value, the best doctor, the best in anything, just name it.

Nature as source and model of learning can best be understood as resulting from the human mediation of the creatively given through a consistent application of human thought. Thus, a people's technology is not just a record of their practical modes of production and the accompanying capital but also a history of their thought. As James Redfield (1993) has observed, 'History is not just the evolution of technology. It is the evolution of thought.' For the traditional Igbo, thoughts about how to model their life, about how to provide for basic necessities and about how to survive in their environment start with learning from nature.

The appropriation of nature for the advancement of human interest varies across societies and across gender lines. The origins of gender divisions in hunting and gathering societies, where biology fitted men for hunting and women for food gathering and taking care of the domestic front, and with biology making child-bearing the exclusive role of women, conditioned the later development of human societies. Men concerned themselves with the making of such tools as hoes, knives, spears, clubs, guns, security instruments, etc., and women concerned themselves with food, cloth, art and domestic technologies such as palm oil extraction, castor oil extraction, midwifery techniques and medication, childcare techniques and medication, weaving, pottery, body and wall decoration, etc.

There are instances among the Igbo that not only show that the above is generally true but also that there is historical continuity in the persons engaged in these technologies. Perhaps the most telling example is that of Asele, the legendary designer that excelled in the land of the living and the dead. Emeka Agbay (2003) gives us more information on Asele:

> In Igbo (Nimo) mythology, Asele is the being that excelled so well in drawing and painting in Uli medium. Her dexterity was such that she soon outclassed all other artists on earth. Because of the interconnectedness of life on earth and life beyond, she also got to the land of spirits where she performed the same feat, thus emerging as the best artist of all time.

The case of Asele is particularly interesting because it points to many possibilities of what traditional Igbo women must have been in the very distant past. Designing is a highly intellectual activity that falls within the area of fine arts. If women could achieve such mythic proportions in the area of fine arts, then the sky was their limit in applied arts. Pottery, weaving, dyeing cloth, cotton processing, midwifery, etc., were all areas where traditional Igbo women excelled and exercised dominance. The reproductive functions that they carried out, and which kept them largely in the domestic sphere, created the necessity for them to know most intensely the basic requirements for survival within their environment. They were as close observers of nature as the men who went outside the home to hunt and to cultivate crops. It can therefore be said that nature opens itself to man and woman alike, so that each may contribute to the growth of the other. It then appears that only reproduction was biological, given the roles performed by men and women. Indeed, women were more publicly visible in traditional Igbo society than is the case now.

However, the Igbo, like most societies worldwide, have witnessed dynamic changes in their way of life. Among the Igbo, colonialism, which brought along with it new knowledge, new technologies and new values, is the single most important factor that changed the traditional life patterns. These changes affected even the superstructure of Igbo society. Traditionally, a married woman among the Igbo was simply referred to as *nwunye* (mama), but she also retained her maiden name. With the advent of colonialism and Christianity, a Western form of marrying and naming was introduced; the word 'Mrs.' was translated as *oriaku* that is, someone who depends on the husband for her welfare and survival. However, Igbo women showed great dissatisfaction with this, and some argued that they should be addressed as *odoziaku* that is, someone who takes care of wealth. Later, there was an attempt to adjust the name to mean *okpataku*, someone who *creates* wealth. Women were also confused as to what they should call their husbands. Some adopted *oga* or *nnamukwu* both meaning 'master', and this is still largely the practice among many married women in contemporary Igbo society.

These changes to the superstructure were only a reflection of the changed subculture of the society. The Western education that came with colonialism set the pace for this change. Males were the first to be educated, so they were also the first to get paid employment. This meant leaving one's town, local area or even region, as the states were then called. Most employed men took their wives along with them to their places of work. Since many of the women had no education, they had little choice but to stay at home taking care of the children and serving food and dispensing care to their husbands when they came home from work. It was not that they could not acquire skills or engage in productive roles, but they were dislocated from their traditional homes, where their upbringing had been patterned after their social and natural environment. In their natural environment, the women acquired skills that were relevant to their society. Transplanted from this environment, and having little earning capacity (except for a few that became seamstresses, etc.), they led a life completely dependent on their husbands. It was a natural consequence for such women to regard their husbands as *oga* (masters). In the transformed social environments in which many women then found themselves, nature ceased to be a source and model of learning. Imported technologies, imported knowledge and imported values, mostly of a higher-order value, were introduced. Both men and women were naturally disconnected from their roots. Borrowing from Chinua Achebe, it can be said that things fell apart, this time more dramatically and more damagingly for women. The new order made them dependent on men for many reasons. They were either kept out of school or had fewer years of schooling than their male counterparts, they were socialised into new forms of submission and obedience by Christianity, and the English language, through the use of 'he' as a generic pronoun, impressed on females the idea of being subordinate to males. The monopoly of the public realm by males—in politics, in industry, in education, etc.—further heightened this sense of subordination. The early years of colonisation must, therefore, have proved very traumatic to the female population not only in Igboland, but also throughout Africa.

Thus, it was not surprising that women started crying marginalisation and demanding more education for girls. This arose from their justified indignation at the manner in which society had used education to confine women in the domestic sphere. Despite serious efforts by successive Nigerian governments to reduce the gender imbalance in education, a yawning gap continues to exist in educational access in Nigeria. The masculine perspective in education, the preponderance of male teachers, the pervasive emphasis on male perspectives in curricular materials and the insidious influence of culture and religion—all conspire to relegate women in Nigeria to the background as far as access to education is concerned.

In the area of science and technology that is the concern of this paper, a National Workshop by the Women Education Branch of the Federal Ministry of Education was organised in December 1987. Its goal was to promote a greater role by girls and women in science, technology and mathematics for national development. As a result of the workshop, the Nigerian Association of Women in Science, Technology and Mathematics was launched and was 'charged with the responsibility of promoting science, technology and mathematics education among girls and women in Nigeria'. However, in spite of such measures, there is still a very noticeable gender disparity in the ratio of men to women engaged in science and technology studies and employment in Nigeria. For example, fewer girls offer further mathematics at their senior secondary school certificate examination in Nigeria. This further restricts the access of girls to engineering courses, which provide the major technological base in any contemporary society.

It is this restriction in access to science and technology that women and feminists find disturbing. It is equally that the import of epistemic naturalism, of using nature as a source and model for knowledge acquisition becomes important. Nature reveals itself equally to all human beings. People then draw from nature according to their various mental abilities and capacities to implement the insights they get from it. That Igbo women did not practice blacksmithing in the past was not because of any mental disability, but because of the physical exhaustion involved in heating up and hammering metal into form. In any case, the women themselves did not complain about being excluded from that profession, nor was their absence from it held against them. After all, as we saw in the first part of this paper, things exist in twos to complement each other. This complementarity stems from the fact that all things have some intrinsic qualities, which define them essentially and give uniqueness to their mode of existence. In the final part of this paper, we shall see that the identity of beings derives from this uniqueness and that the harmonious manifestation of this uniqueness leads to progress and development in society.

Mathematics, like the kite, is obviously open to everyone, as is the idea of quantitative reasoning, that is, taking stock in terms of the frequencies with which things happen in life and the number of times the happenings affect people. If mathematics is perceived as difficult, it is not because the people who find it difficult cannot reason quantitatively. Rather, mathematics is perceived as difficult because it is constructed in a highly stylised, formal language. In other words, it is the axiomatisation of quantities and frequencies that makes mathematics appear difficult, and this is what removes it from epistemic naturalism.

As we confront the issue of gender, science and technology in Africa, it is not enough to identity the obstacles that prevent the greater entry of women into science and technology courses and professions. Unless there is a revolutionary conceptualisation of mathematics in such a way that the language used to teach and speak it is understandable by everyone, the problem of gender gap in science and technology, especially in the area of engineering sciences, will continue, not only in Africa but everywhere in the world. The language of contemporary mathematics is not a natural language. In the same way, gender inequality is not natural. But suppose it were?

Subjectivities and Gender Relations Among the Igbo

Among the Igbo, it is said that the flute calls a human being in the manner the person behaves. This conveys the idea that one's identity is socially constructed, and that it is largely constructed from what one does. This much is evident in Igbo praise names, which are often titular. Here are some examples:

Ekwueme - One who matches words with action

Omengbeoji - One who gives out when he or she has

Uchichi na egbu okwu - Night that kills a case (used for someone who goes from place to place at night to talk to warring parties and bring about peace between them)

Egenba - Someone undeterred by threats when pursuing a course of action.

Egbeevumbe - 'A kite cannot carry a tortoise' (used for someone who is clever and hardly caught out in actions, etc.)

These examples demonstrate that one is known and identified by what one does. In life, there are many things to be done, things too numerous to be mentioned. People engage in activities in which they are talented, activities in which over time they have acquired expertise. Talent and expertise are relative to persons, history, geography and time. Individual and societal differences therefore exist and are both biologically and socially explainable. Personhood is thereby biologically and socially defined. One is born a male, a female or, occasionally, a hermaphrodite. One also points to oneself according to what one regards as the cumulative attainments of the self at any point in time. Among the Nsukka Igbo, there is this proverb: *omaba siri na -itiyi nkegi m tiyi nke m bu ka-anyi ji aga* (if you put your own and I put my own, that is the way we move forward). This proverb points to a fundamental principle of existence; life is essentially a question of cooperation, a process of walking and working together. This much is evident in the Igbo idea of binary synthesis and complementary dualities, because there is an up if only there is also a down. The world cannot be understood in either/or terms, because if there is no good, then there is no evil, and the good can only exist in contradistinction to that which is not good. It is within this understanding that I want to situate the remaining part of this lecture. In the process, I intend to examine how the proverb mentioned earlier, *onye na nke ya, onye na nke ya* (each person to his or her ability), can inform gender relations in science and technology in Africa. The emphasis of this section will be on the descriptive nature of the process, in other words, on what the proverb implies ought to be the case and in what is supposed to be the norm.

Earlier we posed the question, what if gender inequality is natural? This question has not been posed in order merely to arouse debate. Again, does the statement 'each person to his or her own talent' mean the same thing as 'each woman to a woman's talent and each man to a man's talent'? These questions are not easy to answer because of the dense and complicated nature of traditional Igbo thought and, I should add, of the thought of other cultures too. However, this discussion cannot move forward unless an attempt is made to answer these questions.

First, let me make it clear that the statement 'each person to his or her ability' is not limited by gender perceptions. This is because there is another Igbo proverb which states: *nke nwoke dire nwoke, nke nwanyi diri nwanyi* (let that for man be for man, and the one for woman be for woman). This proverb clearly signals the division of labour by gender, a universal phenomenon, at least of all traditional societies. It is this perception of the division of labour along gender lines that led the Igbo to construct another proverb to the effect that, when a palm tree is felled, a woman climbs it. This arose from the practice whereby women were not allowed to climb palm trees of a certain height. However, with the introduction of the ladder, some women began to use this to climb and cut down palm fruit heads. In other words, inequality in gender capabilities can be overcome by technology and cultural advancement. This is also borne out by current developments in education, for as Francis Fukuyama (1999:12) notes, '[e]ducated, ambitious and talented women broke down boundaries, proved they could succeed at male occupations, and saw their incomes rise'.

If the statement 'each person to his or her ability' is not overtly limited by gender perception, can it, in Irigaray's terms, 'suspend its pretensions to the production of a truth and of meanings that are excessively univocal' (quoted in Freundlieb and Hudson 1993)? Univocality may be a virtue in Western discourse, but consider the following Igbo proverb: 'when a proverb is used, the wise person understands, and when it is used for the uninitiated, the person flies into the bush'. This suggests already that univocality is not a treasured virtue among the Igbo. After all, when masquerades speak, the negative of what they say is what they mean, making it possible for them to easily catch the uninitiated, who take them at face value. In addition, Igbo words are polysemous and, in orthographic representation, can mislead the uninformed. Thus, the understanding of discourse is a privileged phenomenon among the Igbo, and this privileging can only arise from the multivocatily of discourse.

Against this background, the statement 'each person to his or her ability' refers to what biology and the environment have given an individual or enabled the individual to acquire. It suggests the provision of an unfettered opportunity to enable an individual to benefit from his or her biology and environment, in terms of both innate and acquired abilities. It does not affirm equality of persons, nor does it affirm sameness of persons. It also recommends a way of being in the world and contributing to that world, which implies that a world in which every person is allowed to do what he or she has talents for is a harmonious world. Indeed, this is the key idea of the statement; it is also what connects it with the notions of binary synthesis, complementary dualities and epistemic naturalism. Thus, the Igbo may easily agree with M. Esther Harding (1970) that '[t]he first condition for an impartial investigation into the relationship between man and woman is to rule out old assumptions of the superiority or the inferiority of one to the other.'

However, this does not mean that there are no proverbs in Igbo that inscribe women as inferior to men. They are abundant. However, there are also proverbs that depict the awe in which women are held and the mystery that woman is to man (Opata 1992). The question of men writing women and women writing men is a question of power and must be distinguished from the context in which men and women engage in discourse about human beings. This is also the point made by Rosemarie Tong when she writes:

> Is there easily a way to treat women and men differently yet equally without falling into some version of the pernicious 'separate but equal' approach that characterised the official race relations in the United States until the early 1969s? Or must liberal feminists work towards the elimination of differences as the first step toward true equality? If so, should women become like men in order to be equal with men? Or should men become like women in order to be equal with women? Or should both men and women become androgynous, each person combining the correct blend of positive masculine and feminine characteristics in order to be equal with every other person (Tong 1989).

The issue of inequalities in society should not be smudged within a universalising narrative. The five fingers are not equal, as most cultures recognise. There are the rich and the poor, the privileged and the dispossessed, the politically powerful and the politically marginalised, etc. There must be differences in life, a diversity of talents which when pooled together lead to progress and development. Undoubtedly, the major difference between men and women is sex, not gender, since gender is a fluid and shifting human construct. The signaling of this difference is the performance of reproduction by women. No human being chooses what sex he or she wants to be, at least not at the time of birth. The reproduction/mothering function of women is what mainly delays or impedes their equal entry into certain professions, but it is not enough for women to repudiate that function and begin to see it as a curse. In Africa, especially in the area of science and technology, this is true for at least two reasons. First, parental care of the child, especially the role of the mother, is not yet 'paradise lost', even when deep and rugged inroads have been made into that paradise. The tokenism with which the West marks and brackets off 'lost worlds' must not be encouraged in Africa. There is no need, for instance, to devote one day or year to the international day or year for the child in a society where people live their lives *for* the child, in a society where a child is the child of all, at least as the traditional Igbo adage would have it.

Secondly, science and technology have taken roots in Africa, and very strong roots for that matter, but there has been no adequate measurement of their impact on the family. The extended family system is fast disappearing; even the nuclear family is endangered as both men and women have fewer and fewer hours to spend with their families. We do not, in Africa, want the family dead so that we can declare an international day for the family. In 1996, President Clinton had the issue of the family as one of his major platforms for campaign for re-election. In Africa, we do not want things to get so bad that the recovery of the family ideal should become a campaign issue for politicians.

When the products of Western science and technology have wrecked the world with violence and made international peace a distant rumbling of the already-violated sky, an international year for peace will no doubt be declared. How can there be peace in a world where might, created by science and technology, has displaced and killed right?

It would be foolhardy to argue that Africa should reject Western science and technology as a whole. However, there must be a strong dose of African humanism, as encapsulated in the Igbo adage that prescribes a level playing ground for each person to actualise his or her talents. A man should be left to do what he is best at doing, and a woman should also be left to do what she is best at doing. Things exist as male or female, not in binary opposition but in complementary duality. If a male entity stands and there is no female entity that stands by it, then incompleteness is on parade. The same thing applies if a female entity stands when there is no male standing to complement it. Let the kite do what it has talents for and let the eagle pursue its nature too. So long as each human person has the opportunity to pursue what he or she is good at, there will be peace and harmony in the world.

Onye na nke ya, onye na nke ya! What a resolute encapsulation of diversity—the very stuff of life. However, if any person would use the privilege of difference to stunt the development of another person, may the person's wings be broken; in other words, may the person be stripped of the essence of his or her humanity.

References

Agbay, E., 2003, 'Homage to Asele and Art in Nigeria: A Contemplative Discourse', program notes to *Homage to Asele: An Exhibition in Honour of Uche Okeke*, Pendulum Art Gallery, Lagos.

Aniakor, C.,1979, 'Ancestral Legitimating in Igbo Art', in *West African Religion*, Vol.18, nos. 2-3, pp. 13-30.

Burggraf, S. P., 1997, *The Feminine Economy and Economic Man: Reviving the Role of Family in the Post-Industrial Age*, Reading MS: Addison-Westley.

Freundlieb, D. and Hudson, W., 1993, eds., *Reason and Its Other: Rationality in Modern German Philosophy and Culture*, Oxford: Berg.

Fukuyama F., 1999, *The Great Disruption: Human Nature and The Reconstitution of Social Order*, London: Profile Books.

Harding, M. E., 1970, *The Way of All Women*, New York: Harper Colophon.

Opata, D. U., 1992, 'Igbo Attitude to Women: A Study of a Proverb', in R. Granqvist and N. Inyama, eds., *Power and Powerlessness: Woman in West African Orality*, Stock- holm: Umea Papers.

Opata, D. U., 'Gender and Canon Formation in Nigerian Literature in English: A Search for a Usable Past', paper delivered at CODESRIA International Symposium on Canonical Works and Continuing Innovation in African Arts Humanities, Legon, Ghana, 17-19 September, 2003

Opata, D. U., 1998, *Essays on Igbo Worldview*, Nsukka: AP Publishers.

Redfield, J., 1993, *The Celestine Prophecy*, New York: Warner.

Toffler, A., 1999, *The Third Wave*, New York: Barnes and Noble.

Tong, Rosemarie, 1989, *Feminist Thought: A Comprehensive Introduction,* San Francisco: Westview Press.

Umeh, J. A.,1997, *After God Is Dibia: Igbo Cosmology, Healing, Divination and Sacred Science in Nigeria,* London: Karmak House.

Umezinwa, W. and Animalu, A., 1988, *From African Symbols to Physics: The Meaning of the Snake Symbol in the African Novel and the Implications for Modern Physics,* Lagos: Walaka.

Part II

Science and Technology in Education

4

Educational Policies and the Under-Representation of Women in Scientific and Technical Disciplines in Niger

Elisabeth Sherif

Introduction

The issue of insufficient access to education and training in science and technology was the second point in the platform developed by African women at their preparatory conference held in Dakar before the Beijing Conference.[1] Today, more than a decade later, women's access to education remains a concern, and shocking statistics keep this issue among the top priorities, not only for the independence of African women, but also for the goal of sustainable development on the continent.

Thinking about women's access to scientific knowledge in Africa generally follows a deterministic approach that stresses the biological, psychological and social factors affecting girls' behaviour within the educational infrastructure available to them (Erinosho 1994). Using the example of Niger, this chapter will attempt to show how the available school infrastructures, the prevailing issues at the time of their creation and the ideologies underlying girls' access to education affect their choice of an area of specialisation. This study presents the under-representation of girls in schools in general, and in the scientific disciplines in particular, not as a result of their constitution or of some biological inability, but as a social construct (Imam 1997; Harding 1999).

This chapter is organised around two main ideas. The first is that, despite their apparent gender neutrality, educational policies have an impact on girls' access to scientific and technical education. The second is that the under-representation of women in scientific study courses is an obstacle to development; it guarantees the reproduction of the cycle of gender-linked inequality by maintaining women in a situation of socio-economic dependence.

In Niger, the educational system has three levels: primary (elementary school), secondary (middle school and high school) and post-secondary (university and other post-secondary educational facilities). An area of specialisation is only chosen after the BEPC, that is, after a minimum of years of schooling.[2] Analysis of the conditions of women's access to scientific education raises questions regarding the access of girls to basic education as well as their performance (UNESCO 1994; Beoku-Betts 1998; Abder and Mehta 1999; Ka 2001), which respectively involve the phenomena of 'under-schooling' and 'poor schooling' to which they are subjected (Ki-Zerbo 1990).

Indeed, the overall enrolment rate of children in Niger, estimated at 30.3 percent, is very low. As well as being characterised by sizeable gaps between girls and boys on the one hand and rural and urban areas on the other hand. Although women represent 50.4 percent of the 10.8 million inhabitants of Niger, the number of girls in educational facilities is consistently lower than the number of boys. In 1990, for instance, they made up only 36.1 percent of the school population at the primary level. This figure rose to 38.6 percent by 1998 and has reached nearly 40 percent today.[3]

However, the increase in the number of girls in school, so often touted by officials as the culmination of their efforts, is misleading, as it does not provide sufficient information about the alarming trend in the number of girls outside the education system. In 1990, the enrolment rate in the general population of girls between the ages of 6 and 11 was estimated to be at only 14.6 percent, as compared to 28.48 percent for boys in the same age group (UNESCO 1994). Moreover, out of 100 new girl pupils in the first year of primary school, only fifteen continue through to the last year of secondary school (as compared to twenty-nine boys), while only one makes it to the end of a university programme. Moreover, technical education is offered in less than two percent of educational facilities, and girls make up barely a quarter of the students (Hamani 2000).

The study on which this chapter is based attempted to identify the political and institutional causes of this under-representation of girls through a critical analysis of the actions and strategies adopted by the government of Niger in its regulation of the education system. Since the government has in recent years been highly dependent on external financing, our study would not be complete without an analysis of its relations with international actors, such as the World Bank, in the context of the application of structural adjustment plans (Stromquist 1998). Consequently, educational policies will be considered not only as a factor that influences the behaviour of the population towards educational institutions, but also as a product of the interactions of political actors within the government of Niger on the one hand and between them and international financing institutions on the other hand.

Furthermore, by presenting equal access to technical and scientific education as one of the principal means of promoting both the socio-economic integration of women and sustainable development, our study posits knowledge in general as an instrument for the promotion of justice and equality. It also recognises the scientific and technological foundations of development. Thus, beyond the analysis of the conditions of girls' access to scientific and technical knowledge, our study aims to assess the efficiency of scientific and technical education—and its capacity to promote equality and development in Africa—through an examination of scientific and technological research and innovation in the socio-cultural context of the situation of women in African countries. Consequently, from the perspective of 'standpoint theories' (Harding 1991, 1993, 1998; Beoku-Betts 2003), it is important to discover what issues and interests are predominant in the creation, legitimisation and distribution of scientific and technological knowledge. What science and what technology are needed in Africa, and for what type(s) of development? And what role can or should women play in the process?

In order to answer these questions, this study relied on documentary research supported by data gathered in the field in Niger in 1997 and 2003.

In the rest of this chapter, I will first attempt to demonstrate, from a historical perspective, how educational policies from the colonial period up to now have created an environment that dissuades girls from studying science and technology. I will then outline the impact this 'masculinisation' of scientific and technical knowledge has had on the socio-economic and political development of women and their societies.

Colonial Education Policies

The role of colonial education policies in the under-representation of women can be seen in both the relative lack of space given to scientific education in colonial schools at the time of their creation and by the fact that the colonial education system stressed the differences between the sexes.

Lack of Emphasis on Science in Colonial Schools

The education policies instituted by the colonial powers in Africa were not focused on transferring knowledge to colonised peoples to enable them to understand, transform and control their environment but to train managers capable of ensuring the proper running of the colonial administration. As the project of schooling the youth among the colonised peoples was essentially aimed at satisfying colonial objectives, the education system evolved along with the needs of the colonial administration. In the French colonies, for instance, the need for interlocutors and the objectives of assimilation contributed to the development of a highly theoretical education, mainly emphasising the mastery of the French language and an understanding of the history of the colonial power. The first school in French-speaking Africa dates back to 1807, but it was not until 1906 that serious discussions were held regarding technical education (Meunier 2000). Professional schools were developed many years after schools with a general educational focus and remained few in number, reserved for a handful of privileged youths who were restricted to an education aimed at preparing them to work in health care, post offices, telecommunications, farming, public works, etc.

The scientific and technical knowledge introduced into Africa at that time was elementary. Despite this, imported techniques and technologies could be described as modern compared to those that were used during the pre-colonial era. However, once again, they were not so much aimed at the development of the local populations as at the successful exploitation of raw materials. Ill-suited to the social and ecological realities of certain pre-colonial societies, imported technologies imposed themselves by disqualifying local knowledge and the balance which it had attempted to maintain up until then between people and their environment (Shiva 1993). These strategies go a long way to explaining the limited development of scientific and technical education in Africa and the 'schizophrenic dichotomy' (Ki-Zerbo 1990) that can be observed between this type of education and the social, cultural and economic realities of Africa.

Colonial Schools and 'Masculinisation' of Knowledge: The Introduction of 'Girls' Studies'

Traditionally, women played a very important role in the production and distribution of traditional scientific and technical knowledge in Africa. However, in the colonies, the education of young girls was distinctly delayed compared to that of boys. As the colonial powers did not initially see a need to use African women in their colonisation efforts, the first girls' school in French West Africa (*Afrique Occidentale Française* or AOF), founded in 1819, was reserved exclusively for European and mulatto girls (Djibo 2000). In Niger, girls' schools only began to be developed a century later, in 1914. Thus, as Djibo notes:

> ...although Ecole Normale William Ponty, a preparatory school for future African heads of state, which brought together the brightest stars from secondary schools throughout French West Africa and provided training for teachers, interpreters and clerks, was opened in 1910, a women's section was only created in 1939. Similarly, the medical school founded in 1918 only received its first classes of midwives twenty years later (Djibo 2000, my translation).

In addition to the fact that they were only able to enter school much later than boys, girls did not have the opportunity to gain access to the little knowledge their brothers were allowed to acquire in those institutions.

Indeed, the ambition the colonial teachers had for African girls was to make them into model housewives and instruments for the transmission and spread of European culture. The education offered to them was essentially aimed at perfecting their status as colonial wives and mothers. The idea that they might eventually make a contribution to public affairs was only envisaged through the education they provided for their children. Georges Hardy, head of the French West African teaching service from 1912 to 1922, summed up the contents and the purpose of girls' education in the following terms:

> When we put a boy in French school, that is one unit; when we school a girl, it is one unit multiplied by the number of children she will have …. When mothers speak French, their children learn it effortlessly and … French becomes quite literally their mother tongue …. [W]e teach them everyday French, a little math and the metric system, purely for application to their household budgets and everyday purchases, but most of their classroom time is devoted to needlework, cleanliness, childcare, hygiene, and housekeeping, and all those lessons maintain a practical, immediately utilitarian character; they are not aimed at developing formulae nor coordinating precepts, they help them develop nimble fingers, and sustain the children's desire for improvement and joy in accomplishing useful work (Hardy 1917, my translation).

Thus, if colonial schoolteachers had finally realised the importance of schooling girls, girl children were assessed only in terms of their contribution to the achievement of colonial aims, and in the light of the prevailing European notions regarding the role and place of women in society. To a certain extent, the developer of the colonial schools merely imported the gender principles of European education systems. Michèle le Doeuff reports that during that era "girls were subjected to a 'girls' education' at the secondary level—not too much science, no Latin, no Greek, no philosophy, but with classes in sewing and home economics" (Le Doeuff 1998, my translation).

Colonial education was thus based on the sexual division of labour that characterised European societies, which adhered strongly to patriarchal principles: 'men were supposed to take care of the economic and political realms, while women were supposed to be responsible for the private realm—the domestic sphere and reproduction' (Mokin 2000, my translation).

Of course, using colonial policies to provide an explanation for the differences between boys and girls in the school system in general, and girls' distance with respect to scientific and technical subjects in particular, does not totally exclude the existence of a gender-based division of labour in pre-colonial Africa. Indeed, it is important to examine the main pre-colonial structures for the transmission of knowledge in order to determine how they dealt with the issue of educating young girls. Did girls receive the same type of treatment as boys in these structures? How did they react to the institution of the new colonial system and, in particular, to its sexual division? Did they develop resistance strategies or, on the contrary, did they cooperate? Such questions urge us to undertake a brief review of the organisation of social relationships between the sexes during the pre-colonial period as well as the role played by women in the mechanisms for the production and distribution of knowledge at that time.

In certain pre-colonial societies in Niger, women's low public profile had less to do with the passive nature that has often been attributed to African women in general, and often wrongly (Djibo 2000; Ki-Zerbo 2003), than with the subtlety of the institutions through which they participated in public affairs (Sherif 1997). Indeed, women in pre-colonial Hausa and Songhai societies played very important roles, both socially and religiously. In Hausa societies, for instance, the positions of *inna*, *iya* and *saraouniya* enabled the women who occupied them to participate in the management of the politico-religious affairs of the state. Songhai women were also able to assert themselves in their societies by monopolising the secrets of worship and healing rituals. It can be argued that the gender structure of those societies reflected, in many respects, a complementary relationship (Mama 1997). However, the complementary nature of those relationships in no way precludes their hierarchic nature, nor their patriarchal essence. In other words, even though gender relations in those societies did not reflect a relationship of domination, such as we can observe in some other regions at that time, they were certainly far from being egalitarian. Moreover, the distribution of roles within these societies was based on gender differentiation. Women's activities were linked, directly or indirectly, to their biological function of reproduction, and the food conservation techniques and herbal medicine they mastered were linked to their social functions as educators and guardians of the physical and mental well-being of their offspring.

Educational institutions in those societies were not geographically defined. They were, as Ki-Zerbo pointed out (1990), 'schools without walls' located throughout the community.[4] Indeed, the community as a whole worked according to its own values and standards to teach children to become men and women who would work toward the greater good of all. Girls' education in Hausa societies, as in other precolonial African societies, destined them to take an active role in their societies, but mainly in areas touching on the private sphere. Linda R. Day reports on the education received by girls in Bundu schools in the following terms:

> The aim of education for girls under the auspices of the Bundu society has always been to transform girls into women. [...] When girls remained in the Bundu School for up to three years or more, they learned how to fish, cook, weave, spin cotton, dress hair, and make baskets, musical instruments, pots and fishing nets. They learned special songs and dances as well as how to behave within the associational structure of Bundu and other corporate groups that comprised the community. The medicinal use of herbs was another traditional skill taught to Bundu initiates. In general, girls in the Bundu School were taught a variety of skills considered essential for a woman (56).

Methods for socialising girls in accordance with community standards and knowledge did not differ greatly from the ideas conveyed by colonial education policies; they also aimed to prepare girl children for their future roles as wives and mothers, so that they could better ensure the reproduction of their society and preserve its moral integrity. The mistrustful attitude parents adopted towards the colonial educational institutions when they were created, far from being a form of protest against sexist education, reflected their desire to preserve their traditional modes of social organisation. Among the Hausa, the Songhai and many other communities in Niger, and in Africa, the prototype of the ideal woman evokes the image of a fertile woman entirely devoted to her roles as wife and mother, respectful and shy, who stays at home and adheres to the code of honour by preserving her virginity until marriage. The colonial schools' location outside family institutions was perceived as a threat to the existing social and cultural balance. The fear that girls would abandon their obligations on leaving home to study was one of the main causes of parental hostility towards colonial schools and the formal schooling of girls.

Thus, it can be seen that gender differentiation in the transmission of knowledge was not unknown to pre-colonial Africa. However, since the colonial schools were an institution foreign to pre-colonial African social structures, it can be argued that an education free of any connotations linked to biological differences could well have produced neutral behaviour toward the subjects taught. Thus, by opening its doors to boys first and by providing only a 'domestically-oriented' education for girls, colonial schools did much to strengthen and justify gender differentiation. The fact that only boys had access to scientific and technical education built up the perception that scientific and technical studies were the natural preserve of the male gender. This guided parents as they socialised their children within an institution that was new to them. Because scientific and technical subjects were widely perceived as 'boys' subjects', girls were more or less conditioned to turn to subject areas more 'suited to their nature'. The internalisation of these stereotypes during the socialisation process creates in girls a distaste for scientific subjects and a belief that girls are more suited for subjects closer to the roles traditionally devolved to them. This explains why the majority of girls can still be found in the arts and education faculties. In addition, even within scientific fields of study, girls in university are more often enrolled in health sciences and in agronomics than in mathematics, physics, chemistry or biology. For example, in 1988, they only made up 2 percent of all students enrolled in science, compared to 18 percent in education. These figures had risen to 8 percent and 24 percent respectively in 1991.[5]

Furthermore, girls who resist the socialisation process aimed at confining them to 'feminine' areas of study have difficulty gaining acceptance in male-dominated areas. Their abilities are constantly questioned, even by other women. A study conducted using a sample of twenty women in Nigeria found that only two preferred to be operated on by a woman rather than by a man and only four would like to work under a female supervisor. However, eleven out of the twenty would prefer to seek a woman's advice in the event of personal problems. This situation is not typical of Nigeria alone. While it is growing less common in the West, such attitudes can be found in most other places in the world. At independence, African states in general did not succeed in deconstructing this socialisation process that the colonial schools had greatly contributed to structuring. On the contrary, they not only perpetuated these tendencies inherited from colonisation, but also created conditions that promoted their institutionalisation.

Post-Colonial Education Policies

The education policies adopted in Niger during the post-colonial period were greatly influenced by the political choices made following independence. These choices shaped not only the content of policy but also the contexts in which policy was designed and implemented. I will outline some of those political choices in order to discern how they not only crystallised unequal access to scientific and technical education between girls and boys, but also strengthened their reproductive mechanism.

Post-Colonial Political Choices

The building of a nation from the disparate groupings inherited from colonisation was one of the major objectives African states set for themselves at independence. This led to the creation of a massive administration through which post-colonial states aimed to control their populations and integrate them into the development process. As early as 1966, in a book entitled *False Start in Africa* (*L'Afrique noire est mal partie* in the original French), René Dumont denounced this phenomenon, describing administration as the top industry in the Third World. Indeed, the attention focused on the edification and management of the administrative system was prejudicial to the development of the industrial sector. The neglect of this sector, which requires workers with scientific and technical training, explains in part the lack of educational infrastructure devoted to scientific and technical training. Girls have been particularly affected by this situation and still form a tiny minority in institutions that were already very few in number. Thus, of the 199 secondary schools in Niger inventoried in 1997, only two were technical institutions. And, as we can see from the following table, very few girls were enrolled.

Table 1: Girls' Access to Technical Education, 1986-1992

Year	Boys	Girls	% Girls
1985-1986	546	75	11.2
1986-1987	601	69	10.3
1987-1988	729	63	8.0
1988-1989	798	51	6.0
1989-1990	775	75	8.8
1990-1991	775	74	8.7
1991-1992	554	91	10.8

Source: Statistical Yearbook of the National Ministry of Education

The rates of access of girls to technical secondary education are far from showing progress. The regression, or at the very least the stagnation, that can be observed in higher education cannot be dissociated from the political and economic trajectory of the government of Niger. Indeed, the methods of government observed in most African states, i.e., the lack of transparency in the management of public affairs and the concentration or monopolisation not only of power, but also of state resources, in the hands of a single individual or groups of individuals (Médard 1991; Bayart 1989; Jackson and Rosberg 1982), have created a climate that does not promote the success of development policies (Mbembé 1992). The inability of post-colonial African states to generate development has bankrupted their economies and made them dependent on international financial institutions (Mkandawire and Soludo 1999). The economic liberalisation demanded by these institutions has led to the progressive withdrawal of the state from public sectors, including education. This has necessitated the adoption of certain policies aimed at restructuring the education system.

Restructuring the Education System: Structural Adjustment and Girls' Schooling

Among the restructuring measures adopted in Niger, two have had particular repercussions on the schooling of young girls and their access to scientific knowledge. These are the establishment of a competition for entry into the civil service and payment by parents of part of the cost of the schooling of their children. The people only began to feel the impact of these reforms in 1990, with the first class of graduates that did not enter the civil service. Parents seriously began to pay for school supplies around the same time.

Before the reforms, young graduates in Niger systematically entered the civil service. This led the population to view state employment as the main purpose of school education and to consider the schooling of their children as an investment for the parents or the group. By introducing a highly selective contest, which became the only means of obtaining a civil service position upon graduation, the state in effect eroded the credibility of education among the population. The growing number of unemployed graduates has accentuated parents' reticence to enrol their children in an institution they increasingly perceive as a 'factory for unemployed people'. Consequently, as Shona Wynd notes, 'schooling is valued not for the basic skills it provides, but for the jobs that students, and their families, anticipate upon graduation. Decreasing job opportunities contribute to perceptions that the time spent learning to read and write is time better spent at home' (Wynd 1995).

Table 2: Enrolment of School-aged Population

Years	Boys	Girls	Difference Boys/Girls
1980	24. 1	13. 71	10. 43
1990	28. 48	14. 60	13. 80
1993	29. 45	14. 55	14. 90
2000	29. 06	14. 40	14. 66

Source: UNESCO, 'L'Éducation des filles et des femmes: par delà l'accès'.

Table 2 enables us to establish a link between the application of these policies and enrolment rates of girls, which rose from 13.71 in 1980 to 14.6 in 1990, but have been falling ever since. At the same time, the gap between the real rates of access of girls and boys to schooling has continued to broaden. Studies in other countries have shown that the correlation between the incomes and social status of parents and enrolment in school seems to be stronger for girls than for boys (Rathgeber 1999; Biraimah 1987; Lebeau 1994). The case of Niger supports that conclusion; the cost-sharing policy seems to be linked to a decrease in the enrolment rate of girls. It should be noted that this measure was introduced at a time when the purchasing power of the parents, which was already quite low, had fallen further and in an environment where the population in general was, as we mentioned earlier, not very convinced that school education was a good idea for girls. Therefore, when their limited means forced them to make a choice of which children to send to school, parents most often favoured their boys. Girls were generally viewed as 'strangers living temporarily with their own families' (Konaté 1992), and since their schooling no longer led systematically to employment, most families deemed it more economically sound to invest in education for their boys. Table 2 above shows that the rate of enrolment among boys continued to progress until 1993, when the girls' enrolment rate recorded its third decrease in a row. By reducing girls' access to basic education, cost-sharing and competition for entry to the civil service together contributed to widening the pre-existing gap between girls and boys in scientific and technical studies. Since basic education is a necessary prerequisite for scientific studies, girls who lack access to school simultaneously lose all hope of acquiring scientific and technical knowledge.

The impact of the cost-sharing and civil-service policies on girls' representation in scientific fields can also be evaluated through the way these two measures affected the path of girls who were already in school. In higher education, new criteria for awarding bursaries were introduced to make it harder for students with humanities backgrounds to get a bursaries and to encourage students to increase their enrolment in the sciences (Issa-Abdourhamane 2000). Since, in Niger, the majority of girls in secondary school enrol specialise in humanities, very few qualify for bursaries. Those who do not have a bursary most often have no hope of pursuing their education and so, bowing to social pressure, turn to marriage. Furthermore, due to reduction in the value of the bursaries, even those girls who choose a scientific major in secondary school are increasingly discouraged from undertaking scientific and technical studies, since the cost of education is higher and their chances of succeeding are slim. Since entering the civil service is no longer a certainty, the argument of private sector hiring practices, which discriminate against women, is also frequently used to dissuade girls from undertaking scientific studies. Consequently, the number of girls enrolled in scientific studies continues to drop, as suggested by the table below.

Table 3: Number of Girls Enrolled in Scientific Faculties
at the University of Niamey

School Year	Faculty of Agronomics		Faculty of Science		Faculty of Health Sciences		Total Students Enrolled	No. of Girls	% of Girls
	Girls	Boys	Girls	Boys	Girls	Boys			
2000-2001	49	201	112	1083	345	774	2564	506	19.76
2001-2002	44	247	66	892	335	818	2402	445	18.52
2002-2003	46	253	51	614	293	834	2020	319	15.79

Source: University of Niamey.

Table 3 provides a recent look at women's representation in the scientific faculties at the University of Niamey. The number of girls enrolled in these faculties dropped from 506 during the 2000-2001 university year to 319 in 2002-2003. And yet, over the same period, the number of girls enrolled in the Faculty of Humanities rose from 687 to 1118.[6] This clearly indicates that girls who have completed scientific secondary studies or who have previously studied in a scientific faculty are increasingly enrolling in the Faculty of Letters and Humanities. The drop in the number of girls studying science has serious repercussions for the women of Niger and the country as a whole.

Consequences of the Under-Representation of Women

Scientific and technical knowledge makes it possible to acquire the necessary expertise to make effective use of scientific progress and innovations to improve the living conditions of the population (Okebukola 1995). The fact that girls are increasingly excluded from the process of acquiring scientific and technical knowledge makes it difficult for women to use new technologies. Furthermore, because women cannot do without scientific and technical discoveries, the refusal to expose them to scientific and technical knowledge places them, as Marie le Doeuff has pointed out, in a relationship of indebtedness and dependency towards men as a group (Le Doeuff 1998). Concretely, women's inability or limited ability to use new technologies hinders them from increasing the productivity and profitability of their activities, which would give them the economic weight to win their struggles for survival and socio-political integration.

Unequal access for girls and boys to scientific and technical education also poses the problem of the under-representation of women and their interests in the structures that design and implement scientific and technical knowledge. The low number of women in these structures, not only in Niger but throughout the world, certainly explains the fact that 'measures aimed at taking account of women's needs in the design and assessment of science and technology have never produced tangible results' (Karzanjian 1999, my translation). Indeed, scientific and technical projects, which are mostly designed by men, cannot effectively address issues linked to the physical, emotional and material well-being of women. Thus, as Olaiton (1994) points out:

> If science is about the search for exact knowledge, there is need to obtain exact information about women and their activities in society. At the most reductionist level, getting exact knowledge about oneself cannot legitimately be done better by another person.

The absence of women in scientific research institutions can only lead, as Sandra Harding has said, to 'scientific and technological changes exclusively designed to simplify the lives of men, and consequently unable to generally improve the position of women, nor stimulate sustainable development (Harding 1999, my translation). From this standpoint, the integration of women into the process of acquisition, production and utilisation of scientific knowledge represents an important issue for women and their societies.

In Niger, 80 percent of women work in agriculture. If these women had increased access to scientific and technical training, they would be better able to understand and use new techniques not just to increase their incomesbut also to help their country resolve the chronic food deficit from which it has suffered for several decades. The country's health indicators are also demonstrative of the need to promote women's representation within scientific structures. In 1990, there was only one doctor per 33,000 inhabitants, only 33 percent of pregnancies were medically monitored and only 21 percent of deliveries were attended by a healthcare worker. Life expectancy in Niger, as in many other African countries, is one of the shortest in the world at 45.7 years, and the mortality rate, at around 20 percent, is also one of the highest. In Niger, women's and children's health is among the most vulnerable in the world. This situation, which reflects both the low level of scientific and technological development in the country and women's exclusion from the process of design and implementation of health programmes, cannot be reversed as long as girls do not have equal access to health training and education. Indeed, since the impact of poor sanitary conditions affects women more than any other segment of the population, it would be wiser and more effective to provide them with the necessary knowledge to influence mechanisms for the identification, assessment and resolution of health problems. This would lead not only to demographic change (Kazanjian 1999), but would also promote the spread of scientific values within African societies in general:

> A young female science student of today is a scientist of tomorrow and at the same time a potential mother and teacher. If she imbibes the traits and attributes from learning science meaningfully, she will most likely pass some on to her children (Azeke 1994).

Thus, through their attachment to the local environment and their decisive role in the process of socialisation of children, women can promote the emergence and spread of a scientific and technological culture adapted to the social context in Niger in particular and in Africa in general. This is especially important because the persistence of underdevelopment in Africa can be viewed as a legitimisation of allegations that science is patriarchal in nature and that technology is an instrument for imperialist exploitation and domination (Shiva 1993; Rathgeber 1999), bringing back to the fore the pressing issue of ownership of imported technologies and their ability to generate sustainable development in Africa (Muntemba and Chimedza 1999).

In this regard, African women can serve as the pillars that hold up the bridge linking modern knowledge to local knowledge. Such a link is the only way to guarantee rational and effective use of imported technologies, but in order for women to fully play this role, they need to be considered not only as receivers but also as creators of knowledge (Appleton et al. 1999). This would mean not only making the acquisition of modern knowledge an imperative for little girls, but also valuing and capitalising on the local knowledge already in the hands of women. This approach would enable African countries to adopt new technologies while maintaining their essence, and respecting their dignity and identity.

Conclusion

Relations structured around gender are complex. On the one hand, their form and breadth vary from society to society and from era to era. On the other hand, Africa is a continent rich in cultural, social and political diversity. Consequently, the present study cannot lay claim to completeness. However, it does provide a look at the political and institutional aspects of the issue of gender, science and technology in one country of the African continent, Niger.

Thus, we have seen that the under-representation of women in scientific and technical studies, which originated in colonial education policies and was perpetuated by post-colonial political choices, remains a major obstacle not only to the social, political and economic integration of African women, but also to the emergence of a process of sustainable development on the continent. Indeed, African countries cannot achieve development while ignoring or under-using the potential of the women who make up more than half of their human resources. They also cannot effectively resolve the issues of famine, ignorance, disease, civil war and poverty if they leave women, who are the most vulnerable members of their population and suffer the most from those scourges, on the sidelines of the development process (Forje 2001).

Women have a very important role to play in the reconstruction of Africa. Through the socialisation process, they can pass on to their children—both boys and girls—a thirst for modern knowledge and a desire to develop traditional knowledge. This will not only stimulate their initiative and creativity, but will also facilitate the emergence and adoption of mentalities and technologies that will promote judicious and productive management and processing of their natural resources, which could lead to development by and for Africans. However, if women are to teach their children to shun handouts and take their future into their own hands, we need to create the conditions that will enable them to free themselves from the dependency imposed upon them by the patriarchal and imperialistic order (Gordon 1996). To this end, numerous policies must be developed and supported by awareness campaigns with a view to promoting women's integration at all levels of the process of acquisition and production of knowledge.

If we view science as an activity aimed at understanding nature and improving human lives, we can imagine that science must suffer at the way it is used to promote the interests of one group to the detriment of another, as well as the attempts to standardise it to the advantage of certain regions in the world and the detriment of the others. Modern science is the fruit of the accumulation of various types of knowledge coming from all over the world. It should therefore be viewed as the legacy of all mankind. It should not be monopolised by certain capitalistic and geo-strategic interests, but should serve all humankind, regardless of gender or regional origins. Modern science is born of diversity and, for it to maintain its essence, we must democratise the conditions of access to scientific knowledge and innovations and create conditions that promote the expression, development and integration of various types of knowledge, so that all nations may see themselves reflected in this universal legacy. Women and men, the West and the rest of the world, 'have different strengths and weaknesses ... reflected in the way they see science and technology and solve scientific and technological problems' (Abder and Mehta 1999, my translation). Science must take this reality into account if it is to shed its patriarchal and Eurocentric attitudes, which may be a homage to the laws of the free market but are a disgrace to humanity. Equal integration of women from all regions of the world into the process of acquisition and production of knowledge 'would add not only new subjects, but would also provide an opportunity to re-examine the premises and standards of existing research' (Mokin 2000, my translation).

To return to the specific subject of gender, one question still begs to be answered. Would the disastrous side of science have developed as it has if women had had the opportunity to participate equally in the design and implementation of scientific and technical projects? In light of the proliferation of weapons of mass destruction and the threat they represent for our lives and environment, it can certainly be said that, more than ever before, the scientific and technical world has need of the protective instinct that women have acquired and developed in the domestic sphere.

Equal participation of women in science and technology in Africa and in the world in general is a major issue that requires mobilisation on the familial, national, continental and global scales in order to establish strategies and policies that will contribute to bringing this about.

Notes

1. For an analysis of the regional platforms presented in Beijing, see Annie Labourie-Racapé, *La quatrième conférence mondiale sur les femmes: priorités et enjeux des programm* 'Part I: Science and Technology in Society: Discourse, Perspectives, Practices and Policy es régionaux' in

Genre et développement: des pistes à suivre, documents et manuels du CEPED, n° 5, Dec. 96, pp. 77-95.

2. The Brevet de fin d'étude du premier cycle (BEPC) is obtained after four years of study at middle school.

3. Statistics gathered by the National Ministry of Education, Niamey.

4. However, other societies did have educational structures that could be described as formal. This was the case in communities that had highly structured secret societies for women, which passed on knowledge to their members through clearly established programmes that were dispensed by specially trained individuals. Bundu society, which existed (and still exists) in certain parts of Guinea, Liberia and Sierra Leone, is one of the institutions that Linda R. Day describes as 'classical African preparatory schools'. See Linda R. Day, 'Rites and Reason: Precolonial Education and Its Relevance to the Current Production and Transmission of Knowledge'.

5. Figures from the Ministry of Social Development, Population and the Advancement of Women (MDS/P/PF), Niamey, 1997.

6. Figures gathered at the registrar's office of Université Abdou Moumouni in Niamey, July 2003.

References

Abder, F. and Mehta, A. J., 1999, 'L'Accès à l'éducation pour tous: une priorité pour l'habilitation des femmes', in *L'autre développement: l'égalité des sexes dans la science et la technologie,* Ottawa: CERDI, pp. 213-229.

Appleton, H., Fernandez, M., Hill, L., and Quiroz, C., 1999, 'Reconnaître le savoir indigène et le mettre à profit', in *L'autre développement: l'égalité des sexes dans la science et la technologie,* Ottawa: CERDI, pp. 55-83.

Azeke, T. O., 1994', 'Beyond Promoting Classroom Participation in Science: The Emerging Role of Women in a Developing Country', in S. Y. Erinosho, ed., *Perspectives on Woman in Science and Technology in Nigeria,* Ibadan: Sam Bookman Services.

Bayart, J. F., 1989, *L'État en Afrique: la politique du ventre,* Paris: Fayard.

Beoku-Betts, A. J., 1998, 'Gender and Formal Education in Africa: An Exploration of the Opportunity Structure at the Secondary and Tertiary Levels', in M. Block, J. A. Beoku-Betts, and B. R. Tabachnick, eds., *Woman and Education in Sub-Saharan Africa: Power, Opportunity and Constraints,* Boulder: Lynne Riemer, pp. 157-184.

Beoku-Betts, J. A., 2003, 'Post-colonial Feminist Critique of Science', Lecture Notes, Dakar: CODESRIA Gender Institute.

Biraimah, K. L., 1987, 'Class, Gender and Life Chances: A Nigerian University Case Study', in *Comparative Education Review,* Vol. 31, no. 4.

Day, L. R., 'Rites and Reason: Precolonial Education and Its Relevance to the Current Production and Transmission of Knowledge', in M. Block, J. A. Beoku-Betts, and B. R. Tabachnick, eds., *Woman and Education in Sub-Saharan Africa: Power, Opportunity and Constraints,* Boulder: Lynne Riemer, pp. 49-72.

Djibo, H., 2000, *La participation des femmes africaines à la politique,* Paris: Harmattan.

Dumont, R., 1966, *L'Afrique noire est mal partie*, Paris: Seuil.

Erinosho, S. Y., ed., 1994, *Perspectives on Women in Science and Technology in Nigeria*, Ibadan: Sam Bookman Services.

Forje, J., 2001, 'Mapping New Futures for Gender Participation towards Sustainable Development: Lessons from Africa', in *Futures Research Quarterly*, Vol. 17, no. 1, pp. 49-60.

Gordon, A., 1996, *Transforming Capitalism and Patriarchy: Gender and Development in Africa*, London: Lynne Riemer.

Hamani, A., 2000, *Les femmes et la politique au Niger*, Niamey: NIN.

Harding, S., 1991, *Whose Science, Whose Knowledge? Thinking from Women's Lives*, New York: Cornell University Press.

Harding, S., 1993, *The Racial Economy of Science: Toward a Democratic Future*, Bloomington: Indiana University Press.

Harding, S., 1998, *Is Science Multicultural? Post-colonialism, Feminism and Epistemologies*, Indianapolis: Indiana University Press.

Harding, S., 1999, L'inclusion des femmes: une panacée ? in *L'autre développement: l'égalité des sexes dans la science et la technologie*, Ottawa: CERDI, pp. 315-329.

Hardy, G., 1917, *Une enquête morale: enseignement en A.O.F*, Paris: Armand Collin.

Issa-Abdourhamane, B., 2000, 'Etudier à l'étranger aux frais d'un etat en crise: le cas des étudiants nigériens à l'étranger, in Y. Lebeau and O. Moboladji, eds., *The Dilemma of Post-colonial Universities: Elite Formation and The Restructuring of Higher Education in Sub-Saharan Africa*, Ibadan: IFRA, pp. 267-285.

Jackson, R. H. and C. G. Rosberg, 1982, *Personal Rule Theory in Africa: Prince, Autocrat, Prophet, Tyrant*, San Francisco: University of California Press.

Ka, M., 2001, 'Le genre dans l'éducation en Afrique: aux sources des inégalités entre les sexes', in *Les jeunes africains et la recherche: des potentialités à renforcer*, Dakar: AFARD/AAWORD, pp. 61-72.

Karzanjian, A., 1999, 'Faire ce qu'il faut, et pas seulement comme il faut: cadre décisionnel en matière de technologie', in *L'autre développement: l'égalité des sexes dans la science et la technologie*, Ottawa: CERDI, pp.169-190.

Ki-Zerbo, J., 1990, *Eduquer ou périr: impasses et perspectives africaines*, Dakar: UNESCO-UNICEF, AAWORD.

Ki-Zerbo, J., 2003, *A quand l'Afrique? Interview with René Holenstein*, Nancy: Editions de l'Aube.

Konaté, G., 1992, *Femme rurale dans les systèmes fonciers au Burkina Faso: cas de l'Oudalan, du Sanmatenga et du Zoun-weogo*, Ouagadougou: Ambassade royale des Pays Bas.

Le Doeuff, M., 1998, *Le sexe du savoir*, Paris: Aubier.

Mama, A., 1997, 'Shedding the Masks and Tearing the Veils: Cultural Studies for a Post-colonial Africa', in A. Imam, et al. eds., *Engendering African Social Sciences,* Dakar: CODESRIA, pp. 61-80.

Mbembé, A., 1992, 'Autoritarisme et problème de gouvernement en Afrique sub-saharienne', *Africa Development*, Vol. 17, n° 1, pp. 37-64.

Médard, J. F., 1991, *Etats d'Afrique noire : formation, mécanismes et crises*, Paris: Karthala.

Meunier, O., 2000, *Bilan d'un siècle de politiques éducatives au Niger*, Paris: Harmattan.

Mkandawire, T., and Soludo, C., 1999, *Notre Continent, notre avenir: perspectives africaines sur l'ajustement structurel*, Dakar: CODESRIA.

Mokin, S., 2000, 'Le genre, le public et le privé', in T. Ballemer-cas, V. Mottier and L. Sgier, eds., *Genre et politique: Débats et perspectives*, Paris: Gallimard.

Muntemba, S. and Chimedza R., 1999, 'Les femmes au cœur de la sécurité alimentaire: science et technologie, un atout?', in *L'autre développement: l'égalité des sexes dans la science et la technologie*, Ottawa: CERDI, pp. 83-106.

Okebukola, P. A. O., 1995, *A Policy Framework for Education and Training*, Manzini, Swaziland: Macmillan Boleswa Publications.

Olaiton, W. A., 1994, 'Perspectives on Women in Science and Technology', in S. Y. Erinosho, ed., *Perspectives on Women in Science and Technology in Nigeria*, Ibadan: Sam Bookman Services.

Rathgeber, E. M., 1999, 'Pourquoi l'éducation? Possibilités d'éducation et perspectives de carrière des femmes dans la science, la technique et le génie', in *L'autre développement: l'égalité des sexes dans la science et la technologie*, Ottawa: CERDI, pp. 191-211.

Sherif, E., 1997, *Women and Politics in Niger Republic*, Ibadan: Sam Bookman Services.

Shiva, V., 1993, 'Colonialism and Masculinist Forestry', in Sandra Harding, ed., *The Racial Economy of Science: Toward a Democratic Future*, Bloomington: Indiana University Press, pp. 303-314.

Stromquist, N., 'Agents in Women Education: Some Trends in the African Context', in M. Block, J. A. Beoku-Betts, and B. R. Tabachnick, eds., *Woman and Education in Sub-Saharan Africa: Power, Opportunity and Constraints*, Boulder: Lynne Riemer, pp. 25-46.

Wynd, S., 1995, 'Factors Affecting Girls' Access to Schooling in Niger', final report to ODA Education Division, Ministry of Overseas Development, London.

5

Girls Opting for Science Streams in Benin: Self-Renunciation or Discrimination in the Educational System?

Ghislaine Agonhessou Yaya

Introduction

Discrimination on the basis of gender exists in every activity throughout the world. Social conflicts over equality between women and men increasingly typify our era. In sub-Saharan Africa in general, and in Benin in particular, the percentage of women engaged in the fields of physics, mathematics and new technologies is negligible. There is hardly a trace of women in areas of advanced specialisation. This is a global phenomenon to such an extent that many people follow *Dhavernas-Levy*, a philosopher at the Centre national de la recherche scientifique (CNRS) in Paris, in asking 'Does science have a sex?' This question leads to two further questions: what exactly is science, and what are the power relationships that exist within our society that maintain science as a male preserve?

In Benin, the absence of women in scientific fields reflects a real social problem that extends far beyond a question of principle. It is an important social and economic issue in a world where the pace of technological change is extremely rapid. Too many people are excluded from the process of taking major decisions about future objectives, most of all women. In successive educational policies, unequal access to scientific education, based on sex, has been criticised as being responsible for keeping women confined to inferior social levels, but the various educational systems adopted in recent decades have not been able to develop any real strategies for democratising knowledge and sharing control of science and technology with women.

One of the principal consequences of this failure is that there are whole scientific and technological areas completely controlled by men. In Benin today, for example, the fields of surgery, higher mathematics, chemistry and physics lie completely outside the control of women, even elite women. In the scientific fields where women are present, they are completely submerged by a tidal wave of men, and they find it very hard to have any counter-balancing part in decision-making.

This chapter presents a critical analysis of the impact that educational policies for the education of girls in Benin have had in steering female students into scientific subjects at secondary and university levels. To some extent, we will be drawn into making analytical reflections on the socio-anthropological factors that influence the selection of scientific subjects at school and university.

Historical Background: A Succession of Educational Systems

There have been three major stages of educational policy in Benin. The first was the educational policy from 1960 to 1975 based on the French colonial system. This policy excelled in producing discriminatory models of the backward-looking attitudes and passivity of Benin society under the colonial system. The result was that 95 percent of the educated elite that emerged from this policy were male. The 'New School' *(Ecole nouvelle)* policy followed from 1975 to 1990. This had the virtue of aiming to democratise access to knowledge by encouraging a huge increase in mass education, but it included no particular policy for promoting gender equity. The policy was based on the then-prevailing Marxist-Leninist ideology and focused on equality of opportunity for the children of the poor. The third stage of educational policy in Benin, in progress since the General Educational Assembly *(Etats généraux de l'Education)* in 1990, is based on the concept of Basic Quality Education *(l'Ecole de qualité fondamentale)*. This new policy is targeted specifically at resolving the problems of inequality between the sexes over access to primary education. One of the steps taken (in 1992) to encourage parents to send their female children to school was the exemption of girls from school fees in state nursery and primary schools. Neither the mass equality system promoted by 1975-1990 policy nor the current policy of encouraging the entry of girls into primary schools made any changes to promote gender balance in scientific streams. Recent statistics continue to show imbalances in the educational system in Benin, and girls continue to opt out of science streams in secondary education and at university.

Lack of appreciation of the consequences of gender imbalance in Benin has affected the economic development of the country. The table below gives an eloquent indication of the imbalances to be seen between the two sexes in the fields of economic and political decision-making in the country.

Indeed, Table 1 reveals some deterioration in the level of women's participation in the economic and political life of the country. This is a reflection of the persistence of sexist traditions in present-day African systems for managing human development. The differences are not only due to the dichotomy connected to the levels of children in primary and junior secondary education, but can also be explained by the parameters of the educational system, which

Table 1: Gender in Economic and Political Professions

	Administrators and other senior officers		Technicians and the liberal professionals		Political power holders	
	Men %	Women%	Men%	Women%	Men%	Women%
Economic decision-making	94	6	83.2	16.8		
Representation in parliament					1990:96 1993:93 2000:94	1990:4.5 1993:6 2000:6

Source: Fieldwork.

leads to the choices of options made by the few girls who are ready to defy all kinds of constraints in order to make themselves useful to the nation at the end of their schooling. In sum, an analysis of the imbalances in the system due to sex can be grouped into two basic causes:

• the educational structure and all its component parts (pedagogy, parents and the extent of maternal control, teachers, educational policy, etc.)

• the intrinsic behaviour of the female individuals who abandon hard sciences, in spite of the equal biological potential of males and females, due to discriminatory models and presentations.

Dysfunctional Aspects of Classical and Disaggregated Systems of Guidance

Children's decisions to follow certain streams of study are determined by many different parameters, including the system for providing guidance, the role of parents, the educational establishment, the teachers and the labour market. These different elements rarely work in isolation. They involve interactions that are both complementary and antagonistic.

A diagnostic analysis in this research of the system of guidance of pupils at school, when they move from junior to senior level, with regard to the parameters referred to above, suggests to us that there are several dysfunctional aspects:

• guidance methods

• parents and families

• schools

• teachers

• the labour market.

Guidance Methods

Various discussions with the managers of the Benin educational system show that the process for guiding pupils between the first cycle and the second cycle of the secondary level (the first time the student is faced with a choice of streams to follow at school) is set up according to mechanical considerations, which reveal some basic determining factors. In 90 percent of cases, the determining factor was provided by the notes and comments in school reports that influenced the choices made by the teachers' council.

The guidance given at school on the choice of subjects is a peculiarly compli-cated matter, since its main objective is to reconcile the development of individual ability with the needs of society. It should not simply depend on one or more spe-cialists – however expert they may be – but should take into account various influ-ences on the student. Also, since these influences are varied and subject to constant change, it is essential that before any decision is taken on choice of subjects for future study, these influences, including extra-curricular ones, should be fully con-sidered. Among these extra-curricular influences, the one that seems to us to be absolutely basic and needs to be taken seriously into account, because of the pre-ponderant role it has on subject choice, is the role of parents and families.

Parents and Families

Parents play a vital role in the choices their children make. The child's first contact with society is not at school, but in the family. Every family can be compared to a matrix whose form serves as a model for the child. In other words, it is the parents who, through the upbringing they give their children, provide guidance according to their own wishes and hopes. In Benin, the views of parents are rarely taken into account in the process that leads to the choice of stream to be followed by girls. Of the parents we interviewed, only 2 percent felt their views had been taken into account, despite the fact that 90 percent of parents expressed a preference about the future careers of their children. In most cases, parents who were illiterate were marginalised and listened very passively to the results that were determined by the iron rule of the notes in the school reports.

Schools

The main criterion in schools in Benin is formed by the collection of a dossier of notes based on examination results, to the exclusion of any acknowledgement of the innate abilities of pupils, which cannot be measured in this way. The role of the school in guiding the choices of pupils at the end of the first cycle of secondary level education should not be confined to setting out periodic school reports, on the basis of which the future of the pupils are decided. Otherwise, apart from a handful of the brightest pupils, who could tackle anything, the children are generally com-mitted to a future that they can only work out slowly and with many hesitations, after a series of successes and failures at school, and by the chance results of examinations and competitions. The role of the schools in helping children to make their choices at the end of the first cycle of secondary education ought to consist much more of concerted actions, first, among the teachers and then between the management of the school and the parents.

Teachers

The traditional system followed in Benin involves head teachers consulting the average marks recorded in school reports and assuming they are then capable of making valid decisions on the future courses of study for their pupils. This system is outmoded, as the reports do not always properly reflect pupils' aptitudes, and even less any sexually specific aptitudes. In those cases where the views of other teachers are taken into account, their views are generally based on the outlines laid down by the authorities, and these outlines do not take gender into account. Indeed, what they contain is really very limited, as regards any analysis of aptitudes that may be gender-based; they are confined to such static descriptions as 'inadequate', 'acceptable', 'satisfactory', 'good', 'very good', etc. This traditional system does not allow us to draw up a profile of the academic development of pupils, whether boys or girls, in the light of the constraints they experience in the school and community environments.

For example, boys at school can quickly become adept in mathematical exercises on probability, often based on drawing cards out of the hat or on games of chance, because, in our societies, it is the boys who have the most time to amuse themselves with such games. The same thing applies to exercises in the physical sciences, which largely concern objects in motion, and which are more readily understood by those who play football, traditionally boys.

Over and above the issue of selection procedures, there is a need for teachers to encourage a sense of motivation among their students. It is this that determines their own approach (both moral and pedagogical) to helping them discern pupils' aptitudes with a view to developing and strengthening them. However, according to Robert Brechon (1970), 'aptitude is not the essential qualification for a particular stream of studies or for a career; it is rather a treasure or a deposit that needs to be exploited'. The exploitation of this deposit requires a different person in the form of a careers guidance adviser, whose function is to provide the pupils with information about each area of study, the state of the labour market and the directions that need to be followed, including those that may be risky, for each stream of studies.

The Labour Market

In Benin, as in many developing countries, the state of the labour market has a big influence on any attempt to guide the direction of pupils' studies. Nevertheless, taking this factor into account – and it is clearly the most important one involved – can make it difficult, if not openly embarrassing, to promote initiatives to guide student choice. Many girls in Benin have learned the truth of the saying that 'in today's world, it isn't enough to have good degrees; the main thing is to have a job'. This explains all the various short cuts that characterise the choice of curriculum to the detriment of hard subjects, which lead inevitably to long courses of study and carry the risk of having to repeat a year's schooling *('redoublement')*. However, as Brechon (1970) argues, '[i]f you take account only of the job opportunities, you sacrifice the individual not only to the community or to the state, but to the economy…You alienate people.' Unfortunately, the labour market remains the key indicator of family pressure concerning the choices made by girls.

Our diagnostic analysis of the process of making choices for stundets shows that the current arrangement is not only defective in the means of appraisal, according to gender, but is incoherent and lacks any interaction between those concerned —the pupils, the teachers, the administration, the parents, the community, etc. One cannot demonstrate this in a better way than by case studies, which allow us to follow the behaviour of girls when confronted with scientific subjects, such as mathematics.

Girls and Science Teaching in Benin: The Case of Mathematics

To understand girls' behaviour more clearly as regards scientific subjects, we decided to use a comparative analysis of national statistics of pupils of both sexes in scientific and literary streams. This was followed up by a survey of a sample of students and teachers in four colleges in the towns of Porto Novo and Cotonou.

The figures in Table 2 enable us to conclude that the science streams in Benin remain the preferred area for boys. This shows the weakness of the educational policies designed to promote gender equality in access to scientific courses. The statistics in the table lead to the following conclusions. In the first cycle of secondary education (from the 6th class to the 3rd), the division between the sexes follows the average of about one girl for every two boys over the five-year period. In the second cycle, the ratio widens to about one girl for every five boys in the science stream, as against one girl for every three boys in the literary streams. From this analysis, it is clear that the imbalance in the science streams is not simply the result of the imbalance linked to the overall number of children of both sexes who are at school. If this were the case, then the difference between the two sexes in the second scientific cycle would have been similar to the difference noted at the level of the first cycle. It follows, therefore, that a considerable number of girls who have completed the first cycle then turn to non-scientific streams.

To demonstrate this phenomenon more clearly, we analysed the specific case of mathematics, which provides the best indicator of the interest that girls show in scientific subjects in general. In practice, not many girls in our colleges and *lycées* opt for science courses, in which mathematics predominates, compared with the number of boys who do so. Those girls who do opt for mathematics do not often achieve good grades.

Table 2: Gender in Science Options in Secondary Schools in Benin (1998 to 2002)

Years		1998	1999	2000	2001	2002
First Cycle						
(Grades 6 – 3)	M	89,965	99,457	108,461	118,433	135,814
	F	39,324	42,845	46,899	52,893	61,814
	T	129,289	142,302	155,360	171,326	196,628
Second Cycle (Literary)						
(Series A and B)	M	3,326	3,342	3,604	3,989	4,324
	F	1,215	1,138	1,194	1,304	1,450
	T	4,541	4,480	4,798	5,293	5,774
Second Cycle (Scientific)						
(Series C and D)	M	15,320	16,580	19,153	2 1,052 2	3,328
	F	3,411	3,834	4,405	4,818	5,619
	T	22,142	20,414	23,558	25,870	28,947

Source: Statistics in the Annual Reports (1998–2002) of the Ministry of National Education.

We carried out surveys by means of samples (pupils, teachers and parents) chosen from four colleges of general education (*collèges d'enseignement général*) at Cotonou (CEG Sègbèya and Ste Rita) and at Porto Novo (Lycée Behanzin and CEG Davié). The surveys were conducted by consulting the head teachers about the pupils on their reports from certain classes in the first and second cycles, by administering questionnaires to the pupils themselves that were designed to reveal the main difficulties encountered in mathematics, by talking with some of the mathematics teachers, so as better to understand the poor results shown by girls in this discipline, and finally by interviewing some of the parents in order to get their views on the direction their daughters were following as well as to estimate the importance of parental influence.

These enquiries enabled us to carry out analyses of the various results that were obtained and to understand the reasons that lay behind the poor results. Table 3 below shows the results from following up a group of students from the Sixth Class to the Third Class in the centres where the survey was conducted. It mainly gives information obtained from the head teachers on the reports on the same group of students over a period of four years at the various levels of the first cycle. This table gives the different results for girls and boys in mathematics.

In Table 3, although the percentage of boys obtaining at least an average mark in mathematics was usually higher than that for girls, we can say that the girls worked as well as the boys in the earlier years. We can see, therefore, that girls can work as well as boys, and even better than them, since in the Third Class of the group being followed, 26.82 percent of boys obtained an average mark, while 28.27 percent of girls did so.

Table 3: Gender in Mathematics Among Students in 4 CEGS
in Cotonou and Porto Novo

	Percentage by Sex with an Average Mark in Mathematics	
	Boys	Girls
6th Class	78	59.09
5th Class	44.73	32.51
4th Class	13.51	9.33
3rd Class	26.82	28.27

Source: Fieldwork.

These results demonstrate that, at the start, there is no proof of innate differences between the female and the male brain. It is therefore paradoxical to see how this initial equality of performance breaks down even further during the second cycle, as shown in the results of Table 4 below.

Table 4: Gender in Mathematics from the Second to the Final Class (Series C and D)
in Cotonou and Porto Novo

	Percentage by Sex with an Average Mark in Mathematics	
	Boys	Girls
2nd Class (scientific)	45	29.09
1st Class (scientific)	28.73	8.51
Final Class (scientific)	13.51	3.33

Source: Fieldwork.

It is worth remarking that these performances are characteristic of urban centres, where the educational infrastructure and the availability of qualified teachers contribute to a reduction of inequality. Other comparable studies, which included colleges from rural areas, produced far worse results for girls, but still showed a better performance by girls during the first cycle than the second.

We note from Table 4 that the performance of girls are invariably worse than that of boys. In the course of the survey, we came across whole classes in which not a single girl achieved an average mark in mathematics during the first half year. When those involved were asked for the reasons for this poor performance, they gave many different answers and provided cause for some serious reflection. Thirty percent of the girls thought their poor performance was connected to the attitude of their teachers (the quality of the teaching, sexual harassment, getting behind because of strikes, etc.) Another 30 percent thought it was linked to lack of time, because the pressure of domestic responsibilities cut down the time available for their homework. Only 10 percent of the girls accepted that their poor performance was due to their own idleness and that they could have done better.

The analyses above show that, to start with, girls have as much aptitude for mathematics as boys. In the classes of the first cycle (see Table 3), the differences between them are not significant. When talking to the girls, the boys and their teachers, it was wholly accepted that girls were no less gifted at mathematics than boys and had the same aptitude for study as boys. In order to establish clearly what were the reasons for their poor performance later on, we have to observe closely all the activities of the girl students within the circle of the family and in the town, as well as at school.

The Family

Within the family circle, the girl suffers from the traditional concepts that parents have of girls. It can be seen all too frequently that her domestic duties get in the way of her homework. Domestic duties compete for her time with study and revision at home. The moment she gets back home, she starts on domestic tasks, and as she gets tired, she gives less and less time to her school homework, particularly to mathematics, which needs great concentration.

In some families, when times are hard, girls are neglected, and preference is given to the boys. The family income sometimes cannot support keeping all children at school. Priority is then given to the needs of the boy(s). And for some parents, the main duty of girls is to get a husband and produce children. This confirms the analysis of Claude Salvy and Judith Paley (1969) that parents tend to 'put the professional career of boys before that of girls…For girls, you still hear all too often the good old cry, You'll get married'.

Relationships with Boys

Academic problems are not the only kind of problems that reduce girls' motivation to pursue scientific subjects. At school, most boys prefer to work without girls, because they don't think the girls are prepared to put forth enough effort, particularly in mathematics, a subject where plenty of concentration and steady willingness to work is regarded as essential. Even when the girls are ready to work, the timetable for group studies disadvantages them, since the girl also has to include domestic duties among the work she has to do.

Attitude of Teachers

A different problem can sometimes lead to poor academic performance among girls. This is the attitude of some mathematics and physics teachers, who benefit from their positions to divert girls from their school work by practices that have little to do with education, such as threatening girls with poor reports, if they do not provide them with sexual favours, regarding the role of a teacher as simply a means of making a living, without attaching any particular importance to it, showing little interest in how well pupils do or in ways and means that could encourage pupils to do better work and underestimating the special problems girls face because of gender inequality. Teachers cannot play their full role unless they seek to be aware of the conditions that would help pupils to show interest in their studies as well as those conditions that reduce pupils' motivation.

The Academic Programme

In a work devoted to the cultural aspects of science, Pierre Thuillier (1988) notes that 'vocabulary that gives preference to males enjoys a wide circulation throughout university libraries, in textbooks and in works of popularisation'. The language of textbooks promotes the myth of mathematics as a 'hard' subject, in contrast to perceptions of sweetness and light that are supposed to characterise symbols of femininity.

The Choice of Science Subjects and the Problematic of Gender: Reflections on a Policy of Equity

The well-known theme of 'innate differences' still makes its way into public discussions (even some women hide behind this argument) and into some popular scientific reviews. However, as Catherine Vidal, a neurobiologist and laboratory head at the Pasteur Institute points out, 'there is no scientific proof that can show any innate differences between male and female brains'. Some tests do show differences such as a greater aptitude of boys for spatial coordination, but one can still argue that these are acquired differences. Perhaps, for example, they are due to playing more outdoor games. Sometimes the arguments about hormones are put forward, but no one has ever been able to prove that these make boys more clever and girls more stupid.

How, then, can one explain the results at the end of the 3rd Class that indicate only slight differences between girls and boys in mathematics, and the later results that show such a striking difference? This is the question posed by Christian Baudelot and Roger Establet (1991), after a fascinating survey of schools. There is no simple answer, they argue, since so many factors are involved, many of them subtle ones, that promote inequalities. An analysis of academic results of girls and boys shows that they have the same results in mathematics as long as they have not yet opted for a particular branch of study. When girls opt for the science stream, the figures show that they perform well here and that many girls are wrongly guided towards a literary stream after a purely mechanical judgment of their aptitudes. It does seem that girls have a less clear vision of what the goal of their studies should be. They claim more often than boys to have chosen the direction of their studies according to their personal preferences and not on the grounds of their professional future. They show themselves to be less certain of themselves when they are confronted with mathematics. Where they are of equal ability in the Third Class, a girl will hesitate before choosing to follow a science stream, while a boy has less fear of being able to cope with the difficulties, because of the need for him to justify what is regarded in our societies as his male bravado. The boy can redo an academic year a second time without feeling that this is any problem for him, since he does not experience as much pressure from the community as a girl, who is faced with the problem of early marriage. Aware of having to take up their responsibilities of motherhood at an early age, girls prudently refrain from engaging in scientific 'adventure' and are more concerned to become useful as soon as possible in their active life. It is this difference, linked to the early start of motherhood, that explains why girls choose the streams that need less training, such as management assistant, office worker, middle-ranking technician, nurse, midwife, teacher, etc.

It is generally true in Benin that girls are naturally excused for not doing well. They join in general discussions less, and they readily resign themselves to self-effacement, but this should not justify the tendency of girls to give up the idea of scientific training. The main lesson of our study is the necessity of overcoming these prejudices in order to develop educational policies that bring together all the aspects we have been reviewing (which have nothing to do with innate characteristics) and that are indispensable for carrying out any process of guidance that would promote sexual equality.

The Role of Teachers

Schools, which should by their nature be neutral, give a highly gender-biased image of mathematics, but teachers are mostly unconscious of this. Many Anglo-American researchers have been trying for several years to determine the factors, normally hidden and invisible, that bring on the biases. With the help of tools such as hidden cameras, they have found that both men and women teachers have a tendency to exploit rivalries between girls and boys in the way they run their classes, and that they expect each group to conform to sexual stereotypes. The science teacher thus gives more of his time to boys. The textbooks, along with the problems to be solved, normally talk of matters of masculine interest. The girls are not asked many questions, and when they are questioned, they are often interrupted. The teachers tend to congratulate the girls on their good behaviour or the neatness of their work, but the boys for the accuracy of their reasoning. Marie Duru Bellat (1995) calls these biases the 'hidden curriculum'. The results of our enquiries confirm the reported biases. The prejudices of the teachers merit being corrected. Teaching involves communicating knowledge by means of given activities, and this communication should not be distorted by discrimination of any kind, and even less by any prejudice emanating from those who are conducting these activities.

In addition, children who come from disadvantaged socio-economic conditions do not usually get any special attention from the teacher, whereas children from middle-class backgrounds benefit from the support and encouragement of their parents at home, which in turn encourages the admiration and motivation of their teachers. The worst of all these situations is when prejudice takes the form of sexual discrimination. Some teachers start from the position that their girl students want to make as little effort as possible. This was very clear among some who took part in our study; girls were regarded as a weaker sex who should be spared such hard subjects as physics and mathematics. None of the teachers involved in the review admitted to such prejudices, but these prejudices are subtle and not easy to detect. They are propagated mainly through internalised socio-traditional practices, and the most effective way of eliminating them would be through re-training sessions on how to promote gender-sensitive discussions in such a way as to ensure equity between the sexes. At the same time, it should be noted that the teaching body faces material problems. The huge numbers of children, the low numbers of teachers and the lack of infrastructure in the schools are serious factors that are outside their control and that sap the energy that could be devoted to a system of pedagogy that respects the equality of the sexes.

The Role of Parents

Do girls unconsciously conform to the expectations of their future employers or more to those of their parents? When asked about their preferences for their children's future, 70 percent of parents said that they would like their sons to take up a scientific career, but only 45 percent said the same for their daughters (*Revue Monde Diplomatique* 1997). Parental decisions greatly influence the direction that daughters take at school. Many boys decide to join a mathematical stream, even when they have only a slight aptitude for mathematics, while this was almost a hypothetical choice for girls. According to Catherine Marie (1992, 1996), young men are systematically propelled by those around them to take up scientific courses and stick to them, even to the point of repeating a year, if necessary, in order to achieve their goals. Girls do not enjoy such support and tend to underestimate their abilities in science when they have to decide which stream to follow.

The impact of the family has a complex influence on the choice of streams. The cases we studied are very varied, and choices are based on various arguments, such as opportunities in the labour market, the security of the family, the differences in jobs among members of the same family and the model provided by one of the parents. Only 22 percent of the girls covered by our research in science classes (C and D) had parents whose careers had any connections with science and technology. Forty percent of the girls had parents following occupations with only a little connection with science and technology. In general, the survey showed that girls' ambitions to carry on their studies to university and post-university levels had a correlation with the level of education and the socio-economic levels of their parents. Only 35 percent of girls with working-class or lower-middle-class parents expected to finish their education at the topmost class in school. Nevertheless, the choice of scientific courses was not entirely governed by the presence in the family of a relation who was an engineer, doctor, mathematician, etc. A significant proportion of girls (20 percent) had no one in their family who had a scientific profession, but it should be added that in most of these cases, their families were engaged in business and their socioeconomic level ensured a good background for their children to be successful at school. Although our research confirms the influence of the family unit in determining whether daughters decided to follow a science stream, the complex means by which this decision was taken obliges us to add that the process is the result of an integration of several lines of influence, rather than simply the parental model.

Some parents require their children to choose the science stream, following the pattern of a parent. One said:

> The choice of stream is an awkward problem for us parents who are concerned over the future of our children. I could see that my daughter was interested in literature. But nowadays, everyone agrees that the world has moved on and that everything depends on science. Everyone has got to take into account that to be able to have a good career, you have got to study scientific subjects. That is why, in spite of her potential for literature, I made her choose the D Series.

In this case, the girl was not interested in the science stream, even though her father was a doctor.

Many families dream of having different specialties among their children. Thus, some girls chose the science streams because their parents do *not* want them to take up the same career as themselves or because one of their brothers chose the literary stream. This means that, even if children have the same aptitudes, they are not allowed to develop in the same way. The need to protect the family by having within it a diversity of professions can also influence family decisions.

Some girls choose to follow science streams when their mother, or some other relation, is in a scientific profession. In theory, this is what one might generally expect to happen, but it is not very common, and sometimes it operates against the choice of science streams. Many girls who opt for literary streams do so in imitation of members of the family who are lawyers, magistrates, writers or teachers, in spite of their own scientific talents. Some parents, it must be said, are not interested in the process by which their children opt for one stream or another. For them, the main point is to choose options that will allow their children to find jobs quickly, so that they can become independent and stop being a burden on the parents. These parents do not encourage their children to opt for subjects that call for a long period of study, whether scientific or not.

Finally, some girls say their choice was not influenced by any outside pressure whatsoever, but was freely based on the results of their own performance. Paradoxically, it often happens that girls who have followed science streams at school choose to study law, languages or economics (to keep up some mathematical facility) at university. This point also covers some girls who start off by being tempted to follow a parent, but who later become disillusioned and change subjects either before the baccalaureate or at university.

These various examples explain the numerous exceptions to the rule of following the parental model, and the danger it presents for the possibility of increasing a scientific feminine elite in Benin.

Conclusion

In Benin, the constitution states that '[w]omen are equal in law to men from every point of view, whether political, economic, cultural, social or in the family'. It is disturbing, therefore, that such key development sectors as science and technology should be so imbalanced in this respect. In concentrating attention on the problem of girls opting for science streams, this study has attempted to throw light on some of the confusion that is engendered by gender preconceptions. Among these pre-conceptions are to be found the question of innate differences between the sexes and the influence of parental example. By means of this modest study, based on a precise sample, we have shown that it is necessary to overcome the prejudice resulting from natural assumptions, as all the parameters that explain the imbalance that we have noted have their origins in social factors and assumptions. The principle of the parental example, which many optimists put forward as a way of overcoming the imbalance between the sexes in the scientific streams, can nevertheless be ineffective, as family pressure may be guided by a variety of arguments and opportunity costs. The concerns thus expressed form the boundary of a study that did not set out to analyse questions of opportunity-cost relating to the choice of the science streams. The result might have thrown up other lines of interest, if the research had covered rural areas and if it had been extended to cover universities.

References

Beaudelot, C. and Establet, R., 1991, *Allez les filles!* Paris: Seuil.

Brechon, R., 1970, *La fin des lycées,* Paris: Bernard Grasset.

Bellat, M. D., 1995, 'Filles et garçons à l'école', in *Revue Française de pédagogie Paris,* n° 110.

Catherine, M., 1992, 'Femme et ingénieur la fin d'une incompatibilité?' in *La recherche,* n° 241.

Catherine, M., 1996, 'Femmes et sciences, une équation improbable?' in *Formation et emploi,* n° 55.

Salvy, C. and Paley, J., 1969, *Les métiers féminins,* Paris: Flammarion.

Thuillier, P., 1988, *Les passions du savoir,* Paris: Fayard.

6

Towards Gender Sensitive Counseling in Science and Technology

Olubukola Olakunbi Ojo

Introduction

This chapter discusses the role of gender-sensitive counseling in relation to female retention in science and technology courses. It is based on a study carried out in Osun State, Nigeria, on the availability of viable counseling services in secondary schools to promote female participation in science and technology. While a wealth of research over the past two decades in Africa shows a consistent pattern of under-representation of girls in science and technology subjects and careers (Eshiwani 1983; Obeng 1985; Manthorpe 1982; Ruivo 1987; Jegede 1998: Erinosho 2001), not much has been done to examine counseling services in schools.

Gender inequality in science and technology is a worldwide phenomenon that has defied attempts to unravel it so far (Erinosho 2001). To worsen matters, the gap between male and female interest and performance in science appears to be on the increase in spite of all efforts. In fact, if the issue constitutes a serious problem in developed Western societies, it is far worse in Africa. In this chapter, reference is made to Nigeria not just because it has the largest population in Africa, but also because, in Nigeria, as in other African countries, the issue of girls' under-representation is so persistent. It was reported in 1992, by the Science Teachers Association of Nigeria (STAN), that less than 10 percent of the total enrolment in Nigerian universities for science and technology disciplines were females, only 6 percent of those enrolled in the West African and the Senior Secondary School Certificate Examinations were girls and less than 5 percent of university academic staff in science-related disciplines were females. This is worrisome in relation to the fact that females make up about 60 percent of the country's 100 million inhabitants. Specifically, less than 30 percent of the one million girls in secondary schools take sciences. This is similar to what obtains in almost every country in Africa and may well be worse in countries with fewer resources allocated to education.

According to Adesemowo (1998) guidance and counseling is an indispensable arm of education. In line with this, the Nigerian National Policy on Education (1977, revised 1981, states:

> In view of [the] apparent ignorance of many young people about career prospects and in view of personality maladjustment among school children, career officers and counselors will be appointed into post-primary institutions [paragraph 85 ii].

Counselors play a key role in the process of streaming students either into sciences or arts. Thus, the importance of counselors for career choice is obvious. They can either influence consciously or unconsciously the perception of females about science and technology domains. Hence, the intention of this chapter to focus on the role of counselors in helping girls into sciences.

Previous studies have shown that usually girls start school with positive attitudes towards the sciences but that their interest diminishes as they proceed further (Erinosho 1997). One of the reasons that have been advanced for girls' diminishing interest in science is that they do not experience science activities and skills in the classroom to the extent that boys do (Khale and Lakes 1983). Available evidence shows that, in terms of their instructional strategy and expectations, teachers treat girls differently from boys (Khale and Lakes 1983; Oakes 1990). The research shows that males generally receive more attention from teachers and dominate classroom activities (Goods et. al.1973; Tyack and Hansot 1990) and also get more exposure to appropriate role models (Hills et. al. 1990). Furthermore, available evidence in the Nigerian context shows that girls often learn science in a gender-biased classroom environment, that is, one in which both the overt and the hidden curricula treat boys and girls differently. According to Okeke (1999), the need to address the issue of gender stereotyping in Nigerian schools is widely accepted, but specific, coordinated programmes or projects are yet to be instituted, unlike in Ghana and Botswana (Ghana Science Clinic for Girls and Botswana Road Shows).

This Study

Even though science educators agree that girls need special motivation (especially counseling) to stimulate their interest in science, no study, as far as I can tell, has investigated the availability and effectiveness of such support in enhancing females' participation in science. Therefore, an investigation of counseling in schools is timely. The key questions guiding the study were as follows: If girls are under-represented in science and technology, have counselors contributed to it and, if so, how? What structures and facilities are in place to encourage girls' interest in science and technology? And, finally, to what extent does the training received by counselors prepare them to be gender-sensitive in science and technology counseling activities?

The main aim of this study was to investigate the effectiveness of counseling for promoting a gender-sensitive curriculum in science and technology. Specifically, the objectives of this study were to:

- find out the availability, or otherwise, of viable school counseling services in science and technology
- determine to what extent sexism prevails among school counselors
- assess the programmes put in place for effective science and technology counseling, such as excursions, career days, seminars and so on
- assess the training of counselors as to whether it adequately prepares them for counseling aimed at retention of females in science and technology.

Studies of this nature are long overdue in Africa. If developed countries are still trying to unravel the mystery, Africa has to double its efforts to demystify the non-participation of girls in science and technology. According to Beoku-Betts (2003), since science is part of the larger society in which we live, we need to understand how its historical and geographical evolution and location has interacted with structured systems of inequality, whether based on race, gender, colonialisation or globalisation, to shape particular consequences for people in those societies. Harding (1991) also notes that, in order to answer the question why so few women in some societies are in sciences, a more comprehensive understanding of the constraints and barriers is required than when scientist of European descent are the primary population under consideration. In the first instance, African's problems as related to development issues are compounded by problems of socio-cultural stereotypes. If the socio-cultural issues are to be tackled, it has to start with counseling girls in school, who will then effect change in the society. If change is to be effected in the schools, it has to be done through effective counseling. Thus, a study of this nature may well be the key to the malaise of non-participation of females in science and technology in Africa.

The study focused on counselors who deal with females between 12 and 20 years. This age group is important because it coincides with the development of the sense of gender identity, the internalisation of gender roles and some of life's major decisions such as education and career choices. The decisions taken during this period largely determine future personal development. Furthermore, UN data shows that 29 percent of the world's population was in the age range of 10-24 years in 1990, and that 83 percent of these lived in the developing countries. It has also been observed that, while this age group is a declining proportion of the population in industrialised countries, it is a rapidly growing proportion in developing countries (WHO 1986). In Nigeria, it has been estimated that there are about 26.5 million adolescents, out of whom 13.8 million (about 52 percent) are girls (NDHS 1990). For African countries to develop, therefore, the preparation and nurturing of youths, and especially girls, is essential.

The research undertaken for this study was qualitative, employing descriptive survey as its research design. The population for the study consisted of all schools in lle-lfe in Nigeria. Sixty counselors participated in the study, as most schools are without practising counselors. These counselors represented 90 percent of all the counselors in schools in Osun State at the time of study. The 'snowball' sampling technique was used; each participating counselor referred the researcher to another school counselor. In the end, 38 female and 22 male counselors were interviewed. The content and validity of the interview instrument was ensured with the assistance of experts in the field and the directors of the 2003 Gender Institute at CODESRIA. The researcher interviewed counselors with assistance from the contacts who helped with their identification. They were interviewed on the counseling programmes available in their schools for science and technology in terms of excursions, career day, seminars and so on. The interviews also elicited responses on the counselors' perceptions of female involvement in science and technology subjects and careers. Furthermore, responses were also elicited on the course content of the training counselors were exposed to in the university, especially with regard to its gender-sensitivity. Thirty schools were visited to determine the viability of counseling services in the schools. The data obtained was analysed using descriptive statistics and other qualitative analytical tools.

The research revealed that only 20 schools out of all those visited had a counselor's office. Indeed, counseling services were poor in general. All counselors also had clerical and administrative duties. Ten of the counselors had postgraduate degrees and fifteen were graduates. Five had no counseling qualifications at all, which is contrary to the requirements stated in the Nigerian National Policy on Education (1981). No school made provision for Guidance and Counseling on the school's timetable. This meant that students who wished to see the counselor could only do so during break time or a free period. Most counselors only addressed the students every last Friday of each month, for thirty minutes, during extra-curricular activities time.

The sixty counselors interviewed were asked their views on girls' participation in science and technology. Some of the responses are presented below:

Response A: 'It is good for a girl to participate in science and technology, but there are certain occupations a woman should not dabble into. Though what a man can do a woman can do, it is better for instance... I cannot imagine a lady going under a vehicle or a car checking out its problems'.

Response B: 'From my own background, there are no differences in gender. My parents did not raise me up or treat me like a girl, or make me conscious of my sex. So, a woman should take part in science and technology'.

Response D: 'In my opinion, it is better for a woman to be careful in choosing science and technology careers, since we have so many professions that she can choose from'.

Response F: 'Girls can be encouraged ... but they have not been. Some engineering courses, like Electrical and Mechanical, that do not involve climbing, girls can do'.

Response H: 'A woman might only be able to cope if the children are not too many or after child-bearing'.

From the above, it is clear that the barriers to female participation in science and technology are subtle. Nevertheless, 90 percent of counselors showed some gender bias. Some tried to hide their bias, but further probing showed they were not free of sexism.

Respondents were also asked about some of the programmes put in place for science and technology. Only nine schools (30 percent) organised career days focusing on science and technology. Three schools (10 percent) invited female speakers and had excursions to science-based and technology-based firms. Only one school distributed any information leaflets.

When counselors were interviewed on their exposure to gender issues during their academic training, some of the responses were very interesting. For example, one reported: 'We were not given gender studies training; it is a new doctrine'. Another said: 'We were exposed, because they made us believe that the problems we are having as women are gender problems'. Another counselor said:

> In the educational sector, people are clamouring that male clients should be counseled by male counselors and vice versa. But in our training, we are advised to counsel anyone we want. We should not see ourselves as running after girls. Even in counseling ethics, your activity with a female client should be professional.

Someone also likened gender studies to psychology and said, 'Yes a little bit, like psychology, childhood and adolescent psychology'. Another response was: 'Yes. Strictly in guidance and counseling; it is part of the training because when you go to the field you are going to work with both males and females. But another counselor said:

> The question of gender does not arise at all in counseling. In guidance and counseling, we refer to every human being in neutral gender. We are not gender-unfriendly or gender-unconscious.

Conversely, this was the response of yet another counselor:

> Not at all, but our association, the Counseling Association of Nigeria (CASSON), is taking that up now. We are moving on from psychological topics to contemporary issues in Nigeria, and gender issue is one of them.

To the researcher, all the responses recorded are similar to the patterns quoted above, and one can infer that the training received by counselors was not gender-sensitive. Most counselors, both male and female, could not effectively describe what gender issues are all about or even state the fundamentals, the ABC, so to speak, of gender studies. Correct differentiation could not be made between gender and sex. None of them understood gender as socially constructed identities located within cultural practices.

In addition, many gendered connotations could be extracted from the counselors' responses, such as the idea that a woman's place is in the home and that a woman is unable to think. Thus, the counselors were sexist without being aware of it. Perhaps it can also be inferred from the study that, coupled with the counseling bias, some gendered identities might have restrained the participation of girls in science and technology. Society has impressed upon girls not to move close to a male, e.g., in religious ideologies. Therefore, if the counselor is a man, the girl child stays away, and vice-versa. The girl child has also been impressed upon to watch what she says, that is, keep silent and avoid talking too much. The implications of this in counseling are that self-disclosure is limited. Effective counseling is thereby hindered, since communication between the girl child and the counselor cannot proceed effectively.

The findings from this study buttress the claims of feminist standpoint theories that women's lives have been erroneously devalued and neglected, yet these lives should be the starting points for scientific research and the generators of evidence for or against knowledge claims. In the first instance, human lives are part of the empirical world that scientists study, but human lives are not homogeneous in a gender-stratified society. Women and men are assigned different kinds of activities in such a society; consequently they lead lives that have significantly different contours and patterns. Indeed, using women's lives as grounds to criticise the dominant knowledge claims that have been based primarily on the lives of men can decrease the partialities and distortions in the picture of nature and social life provided by the natural and social sciences. According to Harding (1991), women are valuable strangers to the social order. Another basic claim for feminist research by standpoint thinkers is women's exclusion from the design and direction of both the social order and the production of knowledge.

Since the responses revealed that counselors had not been exposed to gender studies, some of their sexist attitudes might have resulted from ignorance. In fact, some counselors equated gender studies to courses on counseling ethics, adolescent psychology, child psychology and so on. This might have accounted for their counseling biases and subtle gender discriminations. If counselors were better informed, they could assist in identifying girls that are skilled and talented and can excel in male-dominated fields. Unless the girl child is given information and an opportunity to benefit from unbiased counseling, the vision of equitable participation in science and technology by both sexes will be unattainable. Thus, the assistance of the school counselor is paramount for the involvement of girls in science and technology fields, in order for them to contribute to the development of their own world, totally redeemed from developmental lethargy and prompted to use their brains for development processes.

However, as noted by Malcolm (1993), minority and black women pay a tremendous price for a career in science. For Malcolm, the challenges minority women face in science may the result of factors not only in the scientific community but also in her culture. This statement brings out succinctly the discoveries of this study. Some of the counselors interviewed reflected on the woman scientist's role vis-à-vis the demands of keeping a home and rearing children, which are important issues to a greater percentage of African women. The socially induced need for women always to consider what men or others will think leads to a greater gap between their observable behaviour and speech and their thoughts and judgments.

Conclusion

This study set out to examine the availability of viable counseling services in secondary schools for the promotion of female participation in science and technology. The results reveal that there were no viable counseling services. Moreover, it was found that the counselors themselves were sexist and that there were no programmes on the ground that could assist in promoting girls' participation in science and technology in the schools. Finally, the training programmes experienced by the counselors at university had been devoid of gender studies.

One cannot expect to raise the levels of girls' and women's participation in science and technology without greater understanding of gender and women's studies by counselors. What kinds of interventions can promote participation in science and engineering careers by females? Malcolm (1993) suggests the following: starting early, making appropriate career information available and providing role models who are minority women. There is a need to encourage young girls especially, in order to make sure they participate in sciences from a young age and do not to leave these fields by default. Okeke (1999) argues that the best way to promote girls' access to science is to develop a well-articulated, sustainable and comprehensive programme of sensitisation and training for science and technology educators. Early and appropriate exposure to physical sciences, mathematics and technology is essential.

One of the interventions that counselors can embark on is direct counseling on gender-specific issues. For example, since marriage, child-rearing and work are all essential ingredients in most African women's lives, counselors should ensure that counseling interventions take these concerns into account. Finally, the main recommendations that flow from the findings of this study are the following. Counselor education programmes must include gender studies, practising counselors must be exposed to gender-sensitive training workshops to assist them in their counseling interventions in order to eliminate sexism, and opportunities for early work experience in science-related fields for girls should be encouraged.

References

Adesemowo, P. O., 1998, 'Parents Perception of Guidance and Counseling: An Investigative Study', in *Journal of Research in Counseling Psychology*, Vol. 4, no.1, pp. 14-20.

Beoku-Betts, J. A., 2003, 'Third World Perspective on Science and the Environment', Lecture Notes, Gender Institute 2003.

Eshiwani, G., 1983, 'A Study of Women's Access to Higher Education in Kenya with Special Reference to Mathematics and Science Education', Nairobi: Kenyatta University College.

Erinosho, S. Y., 1997, *Female Participation in Sciences: An Analysis of Secondary School Science Curriculum Materials in Nigeria*, Abridged Research Report no.29, Nairobi: Academy Science Publishers.

Erinosho, S. Y., 2001, 'Gender Science and Technology Participation and Performance in Africa', in V. Reddy, P. Naidoo and M. Savage, eds., *African Forum for Children Literacy in Science and Technology*, Durban: University of Durban Westville.

Goods, T. L., Sikes J. A. and Brophy, J. E., 1973, 'Effect of Teacher Sex and Student Sex on Classroom Interaction', in *Journal of Educational Psychology*, Vol. 65, pp. 74-87.

Harding, S., 1991, *Whose Science? Whose Knowledge? Thinking from Women's Lives*, Ithaca: Cornell University Press.

Hills, O.W ., Pettus, W. C., and Hedin, B. A., 1990, 'Three Studies of Factors Affecting the Attitude of Blacks and Females towards the Pursuit of Science and Science-Related Careers', in *Journal of Research in Science Teaching*, Vol. 27, no. 4, pp. 289–314.

Jegede, O., 1998, 'The Knowledge Base for Learning in Science and Technology Education', in M. Naidoo and P. Savage, eds., *African Science and Technology Education into the New Millennium: Practice, Policy and Priorities,* pp. 151-176.

Khale, J. B. and Lakes, M. K., 1983, 'The Myth of Equality in Science Classrooms', in *Journal of Research in Science Teaching,* Vol. 20, pp. 131-140.

Malcolm, S., 1993, 'Increasing the Participation of Black Women in Science and Technology', in Harding, S., ed., *The Racial Economy of Science: Toward a Democratic Future,* Indianopolis: Indiana University Press.

Manthorpe, C. A., 1982, 'Men's Science, Women's Science or Science? Some Issues Related to the Study of Girls in Science Education', *Studies in Science Education*, Vol. 9, pp. 65-80.

NDHS , 1990, *Nigeria Demographic and Health Survey*, Lagos: Federal Office of Statistics.

Oakes, J., 1990, 'Opportunities, Achievement and Choice: Women and Minority Students in Science and Mathematics', in Cazden, B., ed., *Review of Research in Education,* Vol. 16, pp. 153-222.

Obeng, J., 1985, 'Technology Education and Girls' Participation in Ghana', in *Girls and Science and Technology,* Vol. 2.

Okeke, E. A. C., 1999, 'Scientific, Technical and Vocational Education of Girls in Nigeria', in *Access of Girls and Women to Scientific, Technical and Vocational Education in Africa*, Dakar: UNESCO.

Ruivo, B., 1987, 'The Interesting Labour Market in Developed and Developing Countries: Women's Representation in Scientific Research', in *International Journal of Science Education,* Vol. 15, pp. 385-397.

Tyack, D. and Hansot, E., 1990, *Learning Together: A History of Co-education in American Public Schools*, New Haven: Yale University Press.

WHO, 1986, *Young People's Health: A Challenge for Society*, Geneva: World Health Organisation.

7

Early Scientists Were Men; So Are Today's: Perceptions of Science and Technology Among Secondary School Students in Kenya

Kenneth O. Nyangena

Introduction

Science plays a very important role in our lives and is expected to play an increasingly important role in the future of African countries. As such, it deserves to be represented in all its facets by every stakeholder in society. Since the political independence of most African countries in the 1950s and 60s, education has been considered as both a basic human right and investment in human resources for socio-economic development. With regard to gender, policy-makers have emphasised a commitment to equal access. Yet, in order for women to participate in the production of knowledge, it is important to address not only the question of equal access, but also women's success and equal representation at different levels of the system, including higher education.

The 52 percent of the Kenyan population that is female is still under-represented in scientific and technological fields, yet women's role in family responsibilities, e.g., the provision of food, water and health care is strongly underscored by society (Changeiywo 2001). There are many theories that try to explain the continued under-representation of women in science and technology. The images of men as early inventors of science and technology as reflected in science textbooks, teacher's attitudes towards girls who have an interest in science and technology and the social roles assigned to girls and boys by parents are among the factors that, in general, promote the idea that a girl's domain is in the home and that girls have little capacity for science subjects. It is therefore necessary to systematically and empirically establish students' (and especially girls') perceptions of science and technology, in and outside the school, and how these perceptions are likely to influence their future career choices.

Societal Concepts of Gender Values and Practices

This study was based on the theory of gender as social construct. For centuries, it was believed that the different characteristics women and men exhibited were natural and immutable, determined by biological differences or divinely ordained. These characteristics included both ideas about what was masculine and what was feminine and sets of behaviour considered to be inherently masculine or feminine (e.g., women washing dishes and men working with machines). These perceived characteristics have been shaped and constructed by society. 'Gender' is therefore used to describe socially determined characteristics (e.g., men are rational and women emotional), while 'sex' should be reserved to refer to those characteristics that are biologically determined. It is, therefore, people in a given society who define certain characteristics as masculine or feminine, certain activities as appropriate for men and women and certain norms for relationships between men and women. Prevailing Western conceptions of rationality and objectivity, in spite of their great diversity throughout history, evidently have always been part of the construction of distinctive forms of masculinity. That is, what counts as rationality or objectivity is only what can be given a masculine meaning. Men's preferred styles of reasoning, or standards for maximizing objectivity, have thus come to count as rationality *per se*, leaving women's typical styles and standards marked as infantile.

The conditions of women and men's daily lives and their relative position within their societies are therefore embedded in social, cultural, political and economic frameworks and institutions. However, gender relations and identities are not universal; they vary from culture to culture and sometimes from community to community. For example, it has been reported that villagers in a remote area of Southern Sudan once refused to believe that a Western development worker was a woman because she was driving a Landrover! Understanding gender values and practices is basically central to knowing how societies are organised, how they function and the potential for social (and gender) change.

Women's Under-Representation in Science and Technology

There is a worldwide consensus that women are under-represented in the fields of science and technology and that this under-representation is one of the causes of the continued marginalisation of women in the social, economic and political spheres. Currently, it is the desire of every developing country to attain sustainable development based on self-reliance through the application of science and technology (Ogunniyi 1998). Kenya is no exception and, like other African countries, needs to develop, through science and technology education, a human resources capacity for rapid industrialisation that will ensure economic growth and sustainable development. Strengthening the access of girls to science and technology courses is one way of achieving that development goal.

This Study

The study reported on in this chapter examined girls' and boys' perceptions of science and technology both within and outside school environments in Kenya in order to:

- find out who students' mentors or role models were in science and technology
- establish students' awareness of cultural or school stereotypes that influence their choice of science subjects
- determine students' knowledge about indigenous science and how this knowledge may influence their perceptions of science generally.

The research was conducted in Nakuru, a district 157 km northwest of Nairobi, the capital city of Kenya. The district has five national schools and a number of provincial and district schools. The study used purposive sampling to identify two high schools, one national and one provincial. These two schools were selected because students in both performed very well in science subjects at national level examinations.

A representative sample of 52 students was randomly selected for interviews. At Bahati Girls School, a total of 28 students in Forms Two and Three were sampled, while 24 students in the same forms were sampled at Nakuru High School. In Form Two, students who do well in sciences are streamlined into science subjects, while, in Form Three, students should have made a decision to do either science or arts subjects.

The study was carried out through structured interviews that were quite elaborate and sought to elicit students' responses to a number of issues in science and technology. The researcher used a tape recorder, with permission from administration and respondents, in order to record as much information as possible. The use of the tape recorder was especially important in capturing quotations that brought out interesting information.

Boys as 'Tinkerers', Girls as 'Carers'

There is considerable evidence that girls receive less intensive training in science at primary and secondary school levels. Studies in Canada, the United Kingdom and the United States have revealed that girls routinely receive less attention from teachers, who often give answers directly to girls but provide information to boys to enable them to solve problems for themselves (Bowtell 1996). Although girls' science achievement levels frequently equal or exceed those of boys in early primary school, they commonly drop out of science in secondary school, as they come to perceive science as a male-dominated area with very long working hours.

All these studies suggest that, in general, girls have a stronger interest in human and social issues, whereas boys have a stronger interest in 'tinkering', i.e., understanding the mechanical foundations of technology. Since there is agreement that, in developing countries, girls are faced with a variety of factors that conspire to reduce their participation in science and technology, their perceptions and views are very important in designing programmes that will ensure more female participation in science and technology.

A study conducted in Australia on students aged between seven and twelve revealed negative and stereotyped perceptions of scientists (Bowtell 1996). Bowtell argues that a standard measure of a scientist could be derived using the Draw-a-Scientist-Test (DAST) as an indicator of children's stereotyping of scientists. The study shows that children are strongly influenced by the images they see around them at home, at school and in popular culture. The argument advanced here is that when they harbor negative stereotypes of scientists and engineers as hardy and 'weird-looking', they could then reject science as a potential career. Another study carried out in the United States showed that fifth-grade students generally perceived scientists as whites who do their work in some kind of laboratory. Fifty-six percent of the students saw themselves at a desk, either reading a science book or taking notes, and the majority viewed the use of science outside school as an extension of their school experiences. Other researchers contend that gender-stereotypical images of science and technology are developed, reinforced and projected through various agents of socialisation in the society. Such images arise from the attitudes adopted by parents, teachers, friends and society in general. The same images are also developed and reinforced through out-of-school influences such as cartoons, fiction, television programmes, news coverage and many other activities.

Traditionally, the image of engineering has been heavy, dirty and masculine and argues that the science field has been seen as an area dominated by men, with an assumption that, for a woman to succeed in engineering, she must be tough, aggressive and less 'feminine'. Most images of science are fundamentally masculine and lack appeal for girls because they seem to contradict girls' emerging sense of femininity. Easlea (1983) points out that, if achievement in science is seen as masculine, parents and teachers tend to regard scientific ambition as inappropriate for girls, while girls may also see achievement in science and technology as incompatible with their developing femininity. Since competitiveness and objectivity are seen as attributes of science and technology, the image of science hardly attracts women.

At home, many parents continue to promote sex-stereotyped roles, as by dividing household chores between sons and daughters according to traditional patterns (e.g., boys cleaning shoes and windows, taking out rubbish, washing the car, etc. while girls washed up and mended clothes, laid the table and did shopping). The result of such messages from the home environment is that, when girls reflect on the schools' covert messages on science subjects, the home and school messages coincide, overtly encouraging but covertly discouraging. Girls get the message, withdraw from participation in learning and consequently underachieve in sciences.

Study Findings

The majority of the students interviewed both at Nakuru High School and Bahati Girls School had done their Kenya Certificate of Primary Education (KCPE) in rural public schools. At Nakuru High School, a total of 19 students out of the 24 interviewed attended rural public schools, while at Bahati, 10 out of the 28 students attended rural public schools. As regards family background, most of the students indicated that their parents are in formal employment. The Principal of Bahati Girls noted that students whose background is poor struggled to pay full fees in any calendar year and required financial assistance from various organisations such as Jomo Kenyatta Foundation to pay their fees. However, students from such backgrounds performed very well and have great ambition about their future. 'I would like to make my mother a proud woman one day', said a female student of Bahati Girls whose mother is a single parent.

Students said they got motivation in their studies from various sources. They know who their role models are and are determined not to let them down. Although a few students had parents who held high professional positions, most students parents' were either primary schoolteachers or worked in middle-level jobs, yet they said their parents were their best sources of inspiration. 'I would like to be a civil aviation engineer because my father is currently an electrical engineer with Kenya Power and Lighting', said a girl at Bahati. Another student at Bahati Girls said that her father, who worked as a salesman with a wine importing and distributing company, reminded her when she was in primary school that while most girls in the community wanted to pursue teaching as a profession, he wanted her to pursue science at university.

Most of the students rated their fathers as their number-one source of inspiration, followed by teachers and mothers. The professions of the fathers in particular seemed to have motivated them to pursue science subjects. Few students indicated that they get motivation from sisters, brothers or guardians. A student at Nakuru High said that her mother, who operated a small-scale business, did not encourage her to do science but that she took the initiative to do science because of a neighbor who was pursuing an engineering degree programme in a public university. There was clear relationship between levels of education of the parents and the motivation of the students; parents in formal employment tend to motivate and inspire their children to do science more than those in informal employment.

Students were also asked about their communities' perception of girls pursuing science and technology. Out of the total 52 students interviewed from both schools, 12 came from pastoral communities, such as the Maasai, and it appeared that such communities still viewed women as source of wealth and, by extension, do not fully support the education of girls. The students from pastoral communities, except those whose parents are in formal employment, said that their communities' perception of women scientists is very discouraging and compared this perception unfavourably to that of agricultural communities who they said are more encouraging. Female students from pastoral communities contended that even their own parents keep on reminding them of the importance of marriage and family. 'My mother keeps bothering me about my marriage', said a student from the Masaai community studying at Nakuru High School. These constant reminders of gender role expectation distracted the student from concentrating on her studies so much that she sought advice from a female teacher. According to this student, every human being, whether man or woman, has what it takes to do any profession. Challenged on whether subjects such as physics or chemistry are not inherently masculine, the student responded that, if a woman wanted to be as muscular as a man, she only needed to go to the gym.

Students were asked to define 'scientist' and 'technologist'. Despite giving rather loose definitions, almost all of the students said they knew more male than female scientists and technologists. The point here was not to question whether this is a true reflection of the representation of males and females in scientific technical work (although census figures suggest that it is), but to suggest that the students' perceptions of their surroundings helps to foster the association of masculinity with science.

Interestingly, all but one of the students viewed technology in terms of technical work. Asked what type of science and technology they practiced during holidays, most said they repaired radios, operated kerosene pumps, fixed electrical appliances, etc. Only one student, from Bahati Girls, mentioned methods of food preservation as forms of technology she practiced during the holidays. The student highlighted how she preserved the remains of cooked food in order to be consumed the following day.

In many cultures in Africa, schooling is still seen as a necessity for males but a luxury for females. Most students who come from pastoral communities in Kenya, such as the Maasai, still believe that a girl's place is in the home. A male student at Nakuru High School from the Maasai community noted that a girl's education is as good as useless, since she will only end up getting married. The student's argument was based on the Maasai cultural view that woman are a source of wealth achieved through payment of bride price. Such cultures often deny girls education at even the basic level, and even when girls have access to basic education, they must still overcome immense problems as they strive for equity in access to education, leave alone science and technology.

The students interviewed clearly brought out the different roles girls and boys perform at home. Most girls do household work performed by their mothers, while boys engage in masculine jobs performed by fathers. A male student at Nakuru high remarked, 'I do not cook, but I assist my father to repair his car'. He even said that his father could not allow him to cook by virtue of his sex. The student argued that culture does not allow men to cook and that household chores are a domain of women. However, other students, especially girls, felt that Africans had been policed by their cultures for too long. 'I would like to thank my father, in particular, for insisting that everybody in our home, whether son or daughter, share the cooking', said a female student at Nakuru High School. A female student from Bahati Girls revealed that, besides the household chores and academic work she did during holidays, she also had to double up as an academic, spiritual and social adviser to her siblings. The majority of girls reported that they spend between 25 and 60 percent of their time studying during holidays, while almost all the boys reported spending more than 60 percent of their holiday time doing their studies.

Regarding stereotyping by teachers, a number of students said they were aware of teachers with hidden, negative attitudes towards girls interested in science and technology. The majority of girls, especially in Nakuru High School, said they are taught by teachers who do not use gender-sensitive teaching methods but instead use methods that short-change girls. These teachers show more interest in guiding boys than girls. However, because they were aware of this, the students said they were determined and believed that they will make it. A student from among the girls said, 'I can't be distracted; I am determined to become a neurosurgeon'. This is a clear indication of changing attitudes among female students despite the cultural impediments.

Achievement in science is significantly influenced by teachers, particularly at the elementary level, yet teachers often demonstrate a negative attitude towards science, which is then transmitted to students (Fraser 1992). Many students at Nakuru and Bahati identified inappropriate teaching methods as a major factor influencing girls' negative attitudes toward science and technology. Research indicates that girls' attitudes towards science, along with their achievements and experiences in science, decline considerably during high school. Students at both Nakuru and Bahati were aware of the influence of peer pressure, especially in determining what subjects to pursue.

Many teachers, consciously or unconsciously, use gender stereotypes in their lessons. One researcher, reflecting on the experience of being observed in the GIST project, wrote that, if no obvious interest in the subject or topic is displayed, the male teaching staff often flatter the girls or are mildly flirtatious towards them, finding that this is often a successful way of encouraging them. In the same circumstances, male teachers will probably appeal to the boys' competitive instincts. We accept that treating the sexes differently in this way may well be encouraging them to see their roles in the world of science differently but are reluctant to abandon successful teaching techniques.

Another way in which teachers' preconceptions may influence their relationship with girls and boys is in their perception of who is competent and who needs help. Science teachers (like teachers of other subjects) generally interact more with boys than with girls in class. Girls at both Nakuru and Bahati said they fear consulting male science teachers (especially unmarried ones) because of the gossip that then goes around. Unpleasant stories of dating your teacher discourage girls from consulting male teachers.

While interviewing a female student at Bahati Girls, the student said she hated mathematics as a subject yet did well in other science subjects. She revealed that her mother is a single parent and that her father had abandoned the family. While I could not figure out the relationship between the poor performance in mathematics and the single parenthood, the student said that her mathematics teacher resembled and behaved in every manner like her brutal and unkind father. The principal of Bahati Girls was aware of this case and had tried to advise the student in vain. The student said she did not care about failing mathematics because of a teacher who exhibits her father's characteristics.

The study results revealed that most female teachers in both Bahati and Nakuru taught either biology or chemistry, while most if not all male teachers handled physics, mathematics and computer science. Asked whether students viewed male teachers as better role models than female teachers, most students (including 64.3 percent of Bahati Girls students) strongly disagreed, saying that both male and female teachers are human beings with different potentials. However, while 40 out of 52 students interviewed either disagreed or strongly disagreed with the statement that male teachers are better role models, a female student at Nakuru High claimed that male teachers take more time and are more patient when advising students on subjects and career choices than female teachers. She said male teachers are better in identifying the talents of students, and that they also advise and encourage students more than female teachers who, she said, they make careless and personal remarks when students make small mistakes.

Conclusions

I have discussed so far, to a greater or lesser extent, the masculine image of science as abstract, analytic, 'objective' and controlling, developed so that 'natural philosophers could demonstrate their virility by the scientific and technological appropriation of a mechanical earth' (Easlea 1983). At first sight, the suggestion that science is inherently masculine appears radically different from the suggestion that science is simply packaged in a masculine way. But on closer inspection, these views may converge. If school science were made more girl-friendly, centred on girls' interests and ways of working, would this not be one step to creating a feminine science? And if considerable numbers of girls (and boys) emerge from school having learned science in this way, will science as an institution not begin to change?

There is no doubt that developing countries have put a number of mechanisms in place towards addressing the current under-representation of women in science and technology. However, several issues are still being raised as possible explanations for girls' continued under-representation. This study established students' awareness of the negative attitudes some members of society have towards women with an interest in science and technology. The students are also aware of the challenges facing them in achieving their goals as scientists. The students felt that, unless men, who are the majority in decision-making positions, accept the fact that women are equal partners in development and that their position goes beyond the social role of home-maker, the status quo will remain. The study results indicate that student's perceptions definitely shape their attitudes towards science and that positive perceptions will therefore bring positive attitudes towards science and technology.

It is on this basis that there is need to develop policy guidelines to eliminate the gender bias that is rampant in science and technology education. There is also need to provide infrastructure and funding to develop networks of female scientists, educators and even students. Provision of adequate pedagogy, curriculum, facilities, equipment and supplies for teaching girls is the first step towards increasing their participation in science and technology.

References

Bowtell, E., 1996, 'Australian Children of Scientists', in *Primary and Junior Science Journal*, Vol. 12, no. 1.

Changeiywo, J. M., 2001, 'Gender Perspectives in Science and Technology in Kenya', in *Journal of Education and Human Resources*, Vol. 1, no. 1.

Easlea, B., 1983, *Fathering the Unthinkable: Masculinity, Scientists, and the Nuclear Arms Race*, London: Pluto Press.

Fraser, P., 1992. 'How Can Teacher Education Change the Downhill Trend of Science Education?', in *Journal of Science Teacher Education*, Vol. 3, no. 1, pp. 21-26.

Mwangi, J. G., 2001, 'Using Pre-schools to Reduce Gender Imbalance among Science Professionals in sub-Saharan Africa: Critical Issues', in *Journal of Education and Human Resources,* Vol.1, no. 1.

Ogunniyi, M. B., 1998, 'Effects of Science and Technology on Traditional Beliefs and Cultures,' in M. B. Ogunniyi, ed., *Promoting Public Understanding of Science and Technology in Southern Africa*, Cape Town: Wynland.

8

Looking Beyond Access: A Case Study of Science and Technology Education for Girls in Murang'a District, Kenya

Mweru Mwingi

Introduction

Generally speaking, improvements have been made in education in many sub-Saharan African countries, including Kenya. However, while access to and survival in school have become less of a problem regionally, there are still education inequalities in Kenya, even though it is among the few sub-Saharan African countries renowned for investing heavily in education. Education inequalities are pervasive in Kenya (UNESCO 2004). Many are gender- related and are especially pronounced in the quality of education outcomes, where three features are notable. First, girls' performance in the Kenya Certificate of Secondary Education (KCSE) falls below that of boys with regard to the overall aggregate and science and technical subjects. Second, there is under-representation of females in tertiary education. According to the Economic Survey for 2001 women accounted for only 31.7 percent of students enrolled in the six public universities in 2000/01. In Jomo Kenyatta University of Science and Technology, the only science and technology university in Kenya, only 21 percent were female. The trend was similar in all four national polytechnics (29.2 percent), technical training institutes (39.8 percent) and institutes of technology (43.0 percent). Third, although the quality, relevance and effectiveness of education are mirrored in employment, in 1999/2000 women comprised only 29 percent of waged employees. And except for social services, education services and domestic services, where female representation was slightly over 35 percent, women were less than 20 percent in agriculture and forestry, mining, manufacturing, electrical and water works, building and construction, transport and communications and finance and business.

There is a deeply entrenched gender problem in primary and secondary education in Kenya (UNESCO 2002, 2004), despite the fact that that the 8-4-4 education system introduced in the late 1980s was designed to be practical and technical-oriented (Republic of Kenya 1988:17; 1984:5-8) and to prepare Kenyan youth of both sexes for the technical and scientific workplace[1]. Secondary education is the gateway to tertiary education and ultimately to technical and scientific employment, and so it is important to examine the situation in girls' schools, which are claimed to greatly strengthen female performance and participation in the classroom. Single-sex girls' schools are expected to expand opportunities for girls, particularly in science and technology education (UNESCO 2004). Literature on single-sex schools states that the absence of competitive classmates of the opposite sex promotes learning environments that are free from the gender stereotypes that hold girls back (Marsh 1991) and facilitates engagement that is free from peer pressure and negative competition from boys. In single-sex schools, girls' intellectual curiosity, assertiveness and high self-esteem are promoted. Moreover, the presence of female teachers as role models makes girls see that women can hold power and be strong and confident individuals and leaders who can enter any occupational field. Since the 1980s, numerous studies attest to the academic and psycho-social benefits of an all-female education (Riordan 1985,1990; Whyte et al. 1985; Lee and Marks 1990; Wong et al. 2002). However, with the re-emergence of the co-ed/single-sex debate, there is evidently a need for more critical engagement with girls' schooling and, in particular, the issues surrounding female access, participation and output in science and technology education.

Female Participation in Science and Technology Education

Studies in a number of sub-Saharan African countries show that girls in single-sex schools achieve better grades in Mathematics, science and technology subjects and are more likely to pursue university education and take science and technology courses in higher education than girls in mixed-sex schools (Erinosho 1997; Odaga and Heneveld 1995; Eshiwani 1983a, 1983b). While these benefits are well-known, there are still contradictions in girls' education that suggest that a single-sex education is not entirely a solution in the case of female access to science and technology education. Though science and technology subjects are available to all in secondary education far fewer females take the subjects and more under-achieve at KCSE than males (see Table 1). Regarding this trend, the appropriate question to ask is, 'how accessible are science and technology subjects under the 8-4-4 system of education'?

To adequately respond to this question, it is important to understand the need to deviate from mainstream discourses on education equity, particularly access figures. Table 1 shows that males and females almost equally access science subjects;

the exception is Physics, which has low participation both for males and females. Access figures can be presumptuous, however. According to Gaidzanwa (1999:271), access figures conceal the social, economic and political features that accompany the construction and real nature of disparities. More importantly, they conceal gender, which in Kenya is an important feature of trends and patterns in education. As Kiluva-Ndunda (2001:8) states, 'gender determines the way in which power, property and prestige, educational and employment opportunities are organised, regulated and distributed'. Gender permeates and entrenches itself within education and, if left unexamined, gives an appearance of education equity. Thus, Gaidzanwa (1999:271-272) considers the reassessment of the history and theory of education in Africa to be of critical importance to the examination of gendered patterns.

Table 1: Participation in Science, Mathematics and Technology Subjects by Gender, 2001 KCSE

	Female			Male		
	No. of candidates	%	Mean % pass	No. of candidates	%	Mean % pass
*Mathematics	89,481	46.17	15.83	104,334	53.83	21.20
Biology	85,499	48.30	29.52	91,525	51.70	34.48
Physics	16,225	29.67	22.22	38,425	70.33	26.84
Chemistry	84,534	46.61	29.39	96,862	53.39	23.41
Home Science	10,365	95.17	58.26	526	4.83	51.65
Art and Design	418	35.04	53.64	775	64.96	54.91
Agriculture	44,309	45.45	45.54	53,181	54.55	48.67
Woodwork	24	1.84	51.33	1,277	98.16	50.62
Metalwork	3	0.82	56.00	365	99.18	59.08
Building Construction	46	5.31	39.89	821	94.69	49.31
Power Mechanics	9	2.30	36.77	313	97.70	54.36
Electricity	16	3.22	52.31	481	96.78	54.85
Drawing and Design	93	4.98	25.52	1,774	95.02	42.17
Aviation Technology	0.00	None	43	100.0	60.88	None
Computer Studies	543	48.79	54.44	570	51.21	57.64
Typing with Office Practice	970	95.85	54.15	42	4.15	55.86

Source: Kenya National Examination Council.
*Subject that is compulsory for all.

Framing Issues of Female Participation in Science and Technology

This paper asserts that gender intersects race, class and ethnicity to create a unique set of experiences that give rise to the development of distinctive perspectives. This occurs because being male or female within a social context carries certain connotations in which there is constructed difference based on sex (Sow 1999:45). Where sex is accorded oppressions and privileges, as in education, males and females access and experience education differently because gendering disadvantages females more than it does males at both macro- and micro-structural levels of education.

Given the need to understand that education, and more particularly science and technology education, is situated more broadly in a social, cultural and economic context that extends beyond classroom learning and interactions, this paper takes a feminist standpoint (Harding 1987). Feminist standpoint theory is decisive in extrapolating the causes and reasons for low female participation and achievement in science and technology subjects. Since it focuses on what might be described as distinctly 'female' knowledge (Harding 1991), it constitutes a legitimate way of knowing that is situated specifically in the experiences of women. It accepts that knowledge is socially constructed and interrogates the social processes that construct science and technology education as male. Thus, it provides a means for the analysis of how the history and development of an education system structures and institutionalises gendered frames not only within education but also among school subjects. Standpoint theory takes cognisance of the context and locality, or social location, of knowledge. 'Social location' refers to the way we express the core of our existence in the social and political world (Kirk and Okazawa-Rey 2001:58). It emanates from what is around us and how we are positioned in relation to others, in relation to the dominant culture of our society and in relation to the rest of the world. Social location also determines the kinds of power and privileges we have access to and can exercise, as well as the situations in which we have less power or privilege. In the case of education, it is seen in the development of education and the policies that culminate in the patterns evident in an education system such as that of Kenya today.

Research Methodology

The multiple case studies are case-specific. Nonetheless, they demonstrate macro- and micro-schooling realities and point to what could easily be viewed as typical ramifications of education policy and curriculum implementation in the choice of science and technology subjects in girls' schools. The first level of discussion is a structural analysis of secondary education in Kenya that includes deliberations on curriculum policy guidelines and enrolment in school subjects in form four. The analysis of subject enrollment patterns in single-sex girls' schools in Murang'a district is combined with KCSE results to pose the question of how equally accessible science and technology education really is. The second level of analysis examines schools as institutions. Descriptive statistics drawn from a stratified sample of students in form three and form four in Yellowood, Fort Hall and Dominican (all pseudonyms) illustrate the institutional realities as far as science and technology subjects are concerned. This analysis sets up the foundation for a thematic analysis of individual subject preferences and perceived career futures in science and technology. This micro-level analysis is the third level and uses data drawn from focus-group interviews.

Research Findings and Discussion

The analysis of policy can be divided into two categories, that which determines the effects of policy on different population groups and that which examines content in order to determine the values and assumptions underlying the policy process. Analysis 'for' policy is not entirely different from analysis 'of' policy, except that the latter is more applied. Since policy is an instrument of function with philosophical underpinnings, tracing how gender, science and technology are treated in the 8-4-4 education policy documents reveals how gender, science and technology are constructed at the three levels at which policy operates, the symbolic, regulative and procedural levels[2].

The symbolic is the outer level; it constitutes the vision of the ideal future. As far as science and technology in Kenya are concerned, this future is espoused in *Sessional Paper No. 2 of 1996* (Republic of Kenya 1996), which envisions Kenya as an industrialised nation by 2020. However, with regard to education, a key vehicle in this transformation, the *Master Plan on Education and Training: 1997-2010* advocates the promotion of 'scientific concepts and skills' but is silent on the specifics (Republic of Kenya 1998:198-208) of how this will be done. However, the need for increased relevance and quality in education is cited as a strategic means to improved equity and achievement (Republic of Kenya 1998:72-73). At the regulative level, laws and regulations are designed to put the vision into effect. Education that gives priority to the development of science and technology education, like the 8-4-4 system, is one example (Republic of Kenya 1988). All the same, even with an education system that purports to be scientifically and technically geared, gender equity is still far from being realised in education generally, never mind in science and technology education specfically. Some explanation for this can be found in the place women have in Kenyan legislation generally. Mucai-Kattambo et al. (1995:80) argue that, because patriarchal practices remain unchallenged, many East African women are relegated to the social periphery. According to Kibwana (cited in Ndunda-Kivula 2001:165), 'women's role in Kenya society is fleetingly recognised', and official documents hardly mention the issue of gender equity, a gap that is seen in the peripheral treatment of gender issues in nine education and development policy documents released between 1964 and 1991 (Ndunda-Kivula 2001:55-81). The Ominde Report, the first policy document on education in post-colonial Kenya, is a classic example. It is silent on gender equity except for recommendations that girls' boarding primary schools be established in sparsely populated areas (Republic of

Kenya 1964:65). This silence may be understandable, given the climate of the 1960s, but it set a precedent for future policy that was blind to gender and, in particular, female access to science and technology education (Sifuna 1990:42). The legacy of low female participation and achievement is traceable to the omissions of the Ominde Commission report and others after it. On the ground, it is a regulative and procedural policy problem whose ramifications in education occur in the gender-biased ways that curriculum implementation guidelines are interpreted in schools.

Access to and Participation in Science and Technology Education: The Reality

The foundation of a science- and technology-based economy is in education. Thus, the 8-4-4 system aims to 'provide for a practical oriented curriculum that will offer a wider range of employment opportunities' (Republic of Kenya 1984:1; Republic of Kenya 1988:8). In terms of education output, the aim is to 'ensure that students graduating at every level have some scientific and practical knowledge that can be utilised for either self-employment, salaried employment or further training' (Republic of Kenya 1984:1). In practice, this means that both males and females should have equal access to science and technology education and, more importantly, should both study the subjects up to at least form four. Table 1 shows that such a level of participation is yet to be achieved. Even though the curriculum makes provision for these subjects to both males and females, participation and achievement in science and technology subjects is still gendered. Females appear to have less access to technical subjects. No female has access to Aviation Technology. Percentage scores for females are also lower in all the other science and technology subjects, except Home Science. Interestingly, this pattern is even replicated in girls' schools in Murang'a district, warranting one to question where the achievement advantages in single-sex schools lie.

At university level (see Table 2), female enrollment in the natural sciences, agriculture and veterinary medicine, engineering and architecture, medicine and pharmacy raises questions with regard to whether the pattern is a self-made choice or circumstantial, as males clearinsicaly appear to enjoy better access. To determine whether this pattern is preferential rather than circumstantial, it is important to point out that the implementation of the national curriculum is governed by regulative policies that are regularly updated by the Kenya Institute of Education and that public schools have little autonomy.

Curriculum implementation in public schools in Kenya is standardised. The size of a school (i.e., the number of classes it has) determines how many subjects can be offered in the school. This means that there are restrictions on the variety and kind of subject that a student can study. In many instances, the advantages and disadvantages of school size and, subsequently, of subject variety are linked to the history of the school, its regional location and the quality of its facilities and resources.

Table 2: Public Universities Degree Course Enrolment by Gender for 2002

	No. of Students		Percentage per Programme	
	Male	Female	Male	Female
Education	37,932	19,320	66.3	33.7
Humanities and Social Sciences	7,488	11,405	76.7	23.3
Natural Science	15,037	2,466	85.9	14.1
Agriculture and Veterinary				
Medicine	12,875	1,851	87.4	
Engineering and Architecture	7,974	1,139	87.5	12.5
Medicine and Pharmacy	3,416	837	80.3	19.7
Total	114,722	37,018	75.6	24.4

Source: Universities Joint Admission Board.

Among the 3,000-plus secondary schools in Kenya, 17 are national schools, among which are some of the now-defunct 'high-cost' schools (Republic of Kenya 1988:15). Most of these schools are large, single-sex establishments located in Nairobi or its environs. They have four, five or even more streams and for this reason can offer their students a broad selection of subjects ranging from seventeen to twenty-three (see Table 3 after). These schools tend to attract middle-class children.

A few provincial schools are as large as national schools, but the majority have only three or four streams, and many are also located in rural areas. District schools are more complicated. Many of them fall in the now-defunct 'low cost' category, and some even have *Harambee*[3] school characteristics, which include, among other things, poorly qualified teachers and poor facilities. Most district schools have only one or two streams, so the number of subjects that they can offer is restricted to fifteen (see Table 3). The majority of secondary schools in Kenya are small, with

Table 3: Secondary School Curriculum, Forms 3 and 4

Type of School (Stream)	Core Subjects	Physical Education	Science	Humanities	Technical /Applied Science Subjects	Cultural Subjects/ Foreign Languages	Maximum Number of Subjects
1-2	3	1	3	4	2	2	15
3	3	1	3	4	4	2	17
4	3	1	3	4	4	4	19
5 and above	3	1	3	4	8	4	23

Source: MoES&T, Management of the Primary and Secondary Education Curriculum 2001: Circular INS/ME/A/2/1A/124.

only one or two streams. These schools are scattered across rural Kenya. Many of them are day schools accessible even the children of the poor, because school fees are affordable, although not much else is.

There are substantial returns on economic development when women have secondary education, but even more when they are educated in science and technology subjects. However, the realities of Kenyan education militate against this ideal. The schools with four or more streams are unduly advantaged over those with two or less, as they have a wider choice of subjects to choose from in the technical/applied science category. The schools with few streams serve the majority of Kenyan children, but they are too small to provide access to quality education. For many of these schools, the restrictions are a consequence of social location. The same feature restricts girls from accessing equal educational opportunities with boys. Furthermore, the curriculum implementation guidelines are intended to guide schools to interpret the national curriculum, but instead they create differences among schools, setting students on different paths to adulthood. For girls, there is double disadvantage, first, from the kinds of schools they most attend and, second, from the attitudinal problems known to restrict them from participating in science and technology education. When combined, these two factors effectively relegate girls to 'soft' technical subjects like Agriculture and Home Science.

Girls' Participation in Science and Technology Subjects: A Case Study of Murang'a District

At time of this study in 2002, the most recent statistics on Murang'a district showed that it had a total of 19,422 students in 69 mixed, 7 single-sex boys' and 11 single-sex girls' secondary schools. More girls (10,831) than boys (8,591) were in secondary education, an indication of gender parity (Murang'a District Education Office 2001). 68 percent of girls (7,094) were enrolled in mixed-sex schools and 32 percent (3,737) in single-sex schools. At least 90 percent of all teachers had a university degree or a teaching diploma. However, there were considerably more male teachers (62.45 percent) than female teachers (37.5 percent). There was also a shortfall in science and technology teachers and English teachers (Murang'a District Education Office 2001).

According to KCSE results for 2000, all eleven girls' schools offered three science subjects, Biology, Chemistry and Physics. At least 32 percent (3,737) of the girls in secondary school in Murang'a had the opportunity to study three science subjects. However, the indications are that they did not seize the opportunity. Only five of the eleven schools had girls studying Physics to form four. This pattern shows that the availability of a subject in the school curriculum does not make it accessible. In fact, with the exception of Kiriani, where all form fours studied Physics, the total number of girls taking Physics in the five schools was only 30 percent (225). This pattern is odd, because single-sex schools are associated with achievement advantages, especially for girls. All the same, there are conclusions that can be drawn. First, it appears that girls opt for Biology, Chemistry and Physics rather than

Physical and Biological sciences, which are considered softer sciences, because of the higher value awarded to them in the Kenyan education system. Still, the majority of girls in Yellowood, Dominican and Fort Hall avoid Physics. It appears that being in a single-sex school makes little difference to subject preference.

With regard to subjects in the Applied/Technical category, four of the possible eleven are offered in girls' schools in Murang'a. None of the schools offers wood technology, metal technology, power mechanics, electricity technology, drawing and design technology, building and construction technology or aviation technology. However, these 'hard' technical subjects are found in boys' and mixed-sex schools (Murang'a District Education Office 2000; 2001). The national curriculum has marked them out as boys' subjects. The 'masculine' and 'feminine' construction of school subjects invariably excludes girls from technical subjects. With Home Science and Agriculture as the main technical subjects on offer in all eleven schools, and Computer Studies and Art and Design in two, the full advantages of a single-sex education are limited. The limited variety of technical subjects disadvantages girls with regard to higher education and future careers. The irony of this is that girls in Kenya are expected to compete with boys for courses and jobs that require a background in technology education, yet they do not have ready access to relevant subjects. It is important to note that girls know what is relevant to the job market; for example, only 23 percent (246) study Home Science. Girls in single-sex schools appear to be phasing out the subject, despite its availability and reasonably good KCSE scores.

Patterns of Participation in Science and Technology Education in Three Girls' Schools

High academic achievement among girls has become common in secondary education. Girls' schools have held top positions in the KCSE examinations for the last five years (Aduda 2001). Yellowood has featured among the top hundred schools nationally. It has also held top-school status in Murang'a several times (Murang'a District Education Office 2000). While the high rankings of girls' schools in KCSE examination league tables is an indication that girls have become competitive, it is important to note that only a few girls' schools merit this ranking every year (Okwemba 2001). The more important thing, therefore, is to interrogate whether top performance has any link to participation in science and technology education or to quality outcomes in science and technology subjects.

Murang'a district reveals some interesting contradictions as far as equity and participation in science and technology education are concerned. Although more girls than boys are enrolled in secondary schools, female participation in key science and technology subjects remains low. Participation and performance patterns in the 2000 KCSE were not very different from those at the national level, where only 13 percent (734) of the 5,644 candidates taking 'hard' technical subjects in the 2001 KCSE were female (see Table 1). Single-sex schools had no advantage in subjects in the technical/applied category. In fact, all eleven single-sex girls in Murang'a offered two subjects only, Home Science and Agriculture. Subject elective participation patterns among form threes and fours in Yellowood, Dominican and Fort Hall show Home Science to be a dying subject. Only 19 percent (146) take the subject, a trend that is similar in all eleven girls' schools (see Table 4). It may be thought that girls are breaking away from gender stereotypes linked to this 'feminine' subject, but the growing lack of interest in Home Science is actually due to the fact that girls no longer perceive it as having value in higher education or the job market.

There is a slight difference in science subjects. At 96 and 94 percent for Biology and Chemistry respectively, female participation is nearly equal to that of boys nationally (see Table 1). However, participation was extremely low in Physics (18 percent). The pattern was equally dismal in 2001, when less than 30 percent of the KCSE candidates that took Physics were female (Wassanga 2003:579). This skewed pattern is similar among the eleven girls' schools in Murang'a and also in Yellowood, Fort Hall and Dominican (see Table 4). From this pattern, it can be concluded that access to a school is not enough. Biases abound, and the preference for Biology and Chemistry is evident even in single-sex schools like Yellowood, Fort Hall and Dominican, where three science subjects are offered (see Table 4). The same patterns are reflected in girls' career preferences. Science-related careers such as medicine are preferred to those that require a background in technical education, such as information technology or aviation.

Table 4: Participation in Science and Technology Subjects
by Forms Three and Four Students

School	Maths	Science			Applied Science & Technology			
	Maths	Biology	Physics	Chemistry	Home Science	Agric.	Art & Design	Computer Studies
Yellowood	100 %	82%	26%	100%	9%	11%	3%	none
N=367	(367)	(300)	(97)	(367)	(34)	(40)	(10)	
Fort Hall	100%	100%	18%	90%	18%	27%	16%	16%
N=237	(237)	(237)	(43)	(214)	(42)	(65)	(37)	(37)
Dominican	100%	100%	22%	100%	46%	46%	none	none
N=152	(152)	(152)	(34)	152	(70)	(70)		
Total	100%	90%	23%	96%	19%	23%	6%	5%
(761)	(761)	(689)	(174)	(733)	(146)	(175)	47	(37)

Construction of Science and Technology Careers in Three Girls' Schools

Table 5 shows preference patterns for careers in science and technology among a select sample of form threes and form fours in Yellowood, Fort Hall and Dominican. Medicine is the leading choice, followed by nursing and engineering, but these choices are riddled with contradictions. For example, there is a high preference for nursing (28.9 percent) among girls in Dominican, but it is an induced choice. While all form threes and fours in Dominican take Biology and Chemistry, nursing is preferred not entirely because of philanthropic reasons but because, with low grades and a negative attitude towards science subjects, nursing is perceived as a 'softer' entry point into a medical career. It is an easier bargain as compared to medicine, which is the career of choice in Yellow-wood. Medicine is the focus of career interest in all three schools. It can be attributed to the fact that Biology is a favourite subject among these girls. All form threes and fours in Dominican and Fort Hall take the subject and 302 out of 367 in Yellowood. Women are said to feel encouraged to study science when they know that they will have influence over the uses of advancing technology (Rosser 1993). For many girls from Yellowood, Biology appears to be attractive because, unlike Physics, it is not abstract. Science tends to view the world from a male perspective (Rosser 1993), and this perception clearly filters into school subjects. When scientific theories, practices and approaches are viewed as masculine and used to interpret the natural and physical world, certain perceptions, such as that 'girls cannot normally do well in Physics' or, worse, that 'it is impossible for girls to pass Physics' are encouraged because they are built into science subjects and the science curricula.

Medical careers are prestigious, and the economic returns are conspicuous, especially in rural areas. However, in this case, the influence of role models is also important, as statements such as the following show: 'I admire those who work in this profession'; 'I have admired a family friend who is doing surgery at university'; 'I follow my role model, Ben Carson'.

Pharmacy, Psychology and Veterinary Medicine fall far behind medicine and nursing, and the reason is simple. Dominican girls have no interest in psychology or veterinary medicine as careers. Pharmacists and psychologists are rare in rural areas, while veterinarians are associated with farming, which does not command the prestige of nursing and medicine and is not perceived as a profession that requires education and training like nursing and medicine. The reason for this is lodged in the realities of rural life. Also, the curriculum focuses on commercial agriculture, while women's lived experience of agriculture is small-scale crop production for family food. It is probable that girls fail to reconcile the value of agriculture as a subject with the low economic returns that characterise subsistence farming, which is the reality they see.

The construction of careers related to technology is more complex. As I have already pointed out, the national curriculum tends to restrict girls to the 'soft' technical subjects. Even single-sex girls' schools do not have any advantage in this matter, and as the three girls' schools show, there are few aspirants for careers linked to technical subjects (see Table 5). With the exception of engineering, which straddles both science and technology, there is a poor showing in all the other technical-related careers. Particularly striking is the low preference for computer-related careers. Computer Studies was a relatively new subject in Kenyan schools at the time of this study. The lack of interest in computer-related careers was mirrored in

Table 5: Preferred Careers in Science and Technology Among Forms Three and Four Students (%)

School	Science					Applied Science & Technology									
	Medi-cine	Nur-sing	Pharma-cy	Psycho-logy	Veteri-nary	Architec-ture	Engine-ering	Meteo-rology	Aviation	Com-puter Science	Infor-mation Tech	Agricul-ture	Land Economics	Fashion Design	Interior Design
Yellowwood N=184	32.2	7.1	6.0	1.6	0.0	2.7	11.5	0.5	0.5	0.0	1.1	0.0	0.5	0.5	0.0
Fort Hall N=104	25.4	6.1	3.5	1.8	1.8	0.0	6.1	0.9	0.9	4.4	1.8	0.0	0.9	0.9	0.9
Dominican N=81	22.4	28.9	2.6	0.0	0.0	0.0	1.3	0.0	0.0	0.0	0.0	1.3	0.0	0.0	0.0
Total (369)	28.2	11.3	4.6	1.3	0.8	1.3	7.8	0.3	0.5	1.3	1.1	0.3	0.3	0.5	0.3

the negative attitude towards the subject. In Dominican, where the subject had been newly introduced, girls complained that it was 'hard'. The Computer Studies teacher in this particular school also felt that the subject was too advanced, an attitude that perhaps led girls to believe that they had no future in computers. In Yellowood, orientation to the subject created a desire to study it as an elective in forms four, but in Fort Hall the converse was the case. With an established Computer Studies department, the best in the district, girls in this school perceived a value in the subject, but the competition to secure a place in the small Computer Studies class was discouraging. Many girls therefore opted for other subjects rather than face the disappointment of not being listed among the twenty that could take Computer Studies as an elective subject in Form 3. Finally, Computer Studies is totally delinked from other subjects. Its application as an interactive learning tool that enhances and facilitates computer literacy across the curriculum remains unexplored. Thus, there was a general feeling in all the three schools that Computer Studies was a subject that one could catch up with outside school. Another perception influencing how related careers were chosen lay in the gendered perception of the subject. Girls perceive Computer Studies as a modernised extension of Typing and Office Practice; they see it as a technology that simply enables the ordinary secretary to become a 'computerised' secretary. However, although the future of Computer Studies may therefore seem precarious in Yellowood, Fort Hall and Dominican, there is interest shown among females as indicated by the 48.79 percent female candidature in the 2001 KCSE results (see Table 1).

Aviation Technology is a curious choice of career because the subject is not offered in any of the three schools or any in the district (see Table 1). An interesting finding is that the choice is made mainly on the basis of gender equity. The few girls that had interest indicated that they wanted to penetrate Aviation Technology because it is a male-dominated occupation. Such careers are perceived as fighting stereotypes while also being well-paying and prestigious. The same argument can be extended to careers in meteorology, architecture and even engineering, the most popular in the technical category.

The impact that procedural and regulative polices have on curriculum implementation in schools is visible in subject and career choice patterns. While I have shown that regulative polices do limit subject variety, it is important to point out that, in all three schools, there are classes that are yet to fill up for the subjects offered (see Table 4). Concerning the question of whether a wider variety of subjects would help girls' access the kinds of technical subjects they would enjoy and benefit from in terms of career choice, the answer is no. With the prevailing attitudes in the three schools, a wider variety of subjects is unlikely to benefit girls any more than is already the case, because subjects guide career preferences, and the potential of the subjects on offer has not been fully exploited. To increase the career potential of subjects, science subjects like Physics require attitudinal change if the number of girls studying the subject is to increase. Established technical subjects like Agriculture and Home Science require value to be added to their content and application. Others, such as Art and Design and Computer Studies, require more resources so that more students can be accommodated. If this does not occur, secondary education will remain an education of restricted opportunities as it was for working-class females in the UK (Gaskell 1985). Worse still, the mistaken perception that girls cannot study science *because* they are female (Ndunda and Munby 1991) will continue to prevail.

Conclusion

At the outset of this paper, I stated that social location differentiates schooling. I have argued that, within single-sex schools, girls access and participate in science and technology education differently, partly because of gender bias towards these subjects but also because the allocation of school subjects is not uniform. I have illustrated this using as a case study Murang'a district, where access and participation in science and technology education do not match high female enrollment. In concluding, it will be useful to point out that the problem of female access to science and technology is far bigger than an attitude problem. Regulative policies such as national curriculum guidelines *appear* to equalise access to science and technical subjects, but, as this paper demonstrates, these subjects are not accessible equally to individuals or schools. Although both males and females have access to three science subjects, Biology, Chemistry and Physics, girls do not study Physics. Even those in single-sex girls' schools are not exempted from gender-biased attitudes to the subject. In addition, female access to technical subjects appears to be restricted to 'soft' technical subjects like Home Science and Agriculture, which ironically are conceived as uncompetitive in the job market. Bias towards certain school subjects is also not just a question of preference or of a culture of learning that inhibits full participation and quality achievement. It is a problem with roots in the legacy of education in post-independence Kenya and, particularly, in the historical construction of 'male' and 'female' school subjects (Sifuna 1990) and the dictates of an examination-oriented schooling system where subject choice is pegged to potential examination scores rather then the value of a subject in terms of opportunities for higher education and competitiveness in the job market.

Access to subjects in the applied science/technical category depends on school size. This is a structural barrier that has consequences for the kind of technical subjects that females can access. Schools with four or more streams can offer more subjects than those with only one or two streams. However, the great majority of Kenyan youth live in rural areas, and rural secondary schools tend to be small. Many such schools have two streams or less, so they offer fewer subject options in the science and applied science/technical categories. Ironically, these are the schools that girls access most easily. Concerning the structural barriers that schools create in science and technology education, it would seem that while subjects like Physics remain inaccessible because of attitudinal factors (Wassanga 1997), girls are kept out of technology education because the polices guiding curriculum implementation are gender-blind. The challenges of gender bias in science and technology education are endemic to education sub-Saharan Africa (Beoku-Betts 1998). However, it is important to note that, while the rural location of a school does not prevent the availability of science and technology education, biased regulative policies advantage some schools over others. From this study, there is evidence that the problem of gendered policies shuts out even the best of girls' schools.

Finally, with this kind of trend, it is unlikely that many girls will achieve their dreams of pursuing careers in science and technology. For this reason, a science and technology education has to become an integral part of the education system if Kenya is to realise its goal of becoming an industrialized economy by 2020. Regulative and procedural policies that hinder access to Biology, Chemistry, Physics and all subjects in the applied science/technical category must cease to use gender and school size as criteria for access to these subjects.

Notes

1. The 8-4-4 system of education includes eight years of primary education, four of secondary and four of university. Primary school lays the foundation for a study of science and technology education at the secondary level. Mathematics and science subjects are taught as compulsory subjects in the first two years of secondary school, so there is. However, the quality of this exposure and the selection of related subjects is a critical question.
2. The symbolic level, the regulative level and the procedural level are the three different levels at which policy operates or functions. The symbolic points towards the vision of the ideal future which is the future what policy makers work towards. The regulative level is where policy introduces regulations and rules and laws that should be enforced to assist with reaching of the ideal vision. The procedural level refers to the guidelines and the explanations of who should do what and how it should be done.
3. Harambee schools are community schools. They are a phenomenon of the 1960 and 1970s, when local communities pooled resources and built secondary schools because the demand for secondary school places was higher than the government schools available. Harambee schools were mainly established in rural areas. They are under-resourced, with poorly qualified teachers and often poorly managed day schools.

References

Beoku-Betts, J. A., 1998, 'Gender and Formal Education in Africa: An Exploration of Opportunity Structure at the Secondary and Tertiary Levels', in M. Bloch, J. A. Beoku-Betts, R. M. Tabachnick and B. Roberts, eds., *Women and Education in sub-Saharan Africa: Power, Opportunities and Constraints*, pp. 157-184, London: Lynne Riemer.

Erinosho, S. Y., 1997, 'The Making of Nigerian Women Scientists and Technologists', in *Journal of Career Development*, Vol. 24, no. 1, pp. 71-80.

Eshiwani, G., 1983a. *Who Goes to University in Kenya? A Study of the Social Background of Kenyan Undergraduate Students*, Nairobi: Bureau of Education Research, Kenyatta University.

Eshiwani, G., 1983b, *A Study of Women's Access to Higher Education in Kenya with Special Reference to Mathematics and Science Education,* Nairobi: Bureau of Education Research, Kenyatta University.

Gaidzanwa, R. B., 1999, 'Gender Analysis in the Field of Education: A Zimbabwean Example', in A. Imam, A. Mama and F. Sow, eds. *Engendering African Social Sciences,* Dakar: CODESRIA, pp 31-79.

Gaskell, J., 1985, 'Course Enrolment in the High School: The Perspective of Working-Class Females', in *Sociology of Education,* Vol. 1, pp. 48-59.

Kirk, G. and Okazawa-Rey, M., 2001, *Women's Lives: Multicultural Perspectives,* New York: McGraw-Hill.

Lee, V. E. and Marks, H.M., 1990, 'Sustained Effects of the Single-Sex Secondary School Experience on Attitudes Behaviors and Sex Differences', in *Journal of Educational Psychology,* Vol. 82, no. 3., pp. 588-598.

Marsh, H. W., 1991, 'Public, Catholic Single-Sex and Catholic Co-Educational High Schools: Their Effects on Achievement, Affect and Behaviours', in *American Journal of Education,* Vol. 99, no. 3, pp. 320-356.

Murang'a District Education Office, 2000, *Murang'a District 2000 KCSE Results: A Subject Analysis.* Murang'a: Murang'a District Education Office.

Murang'a District Education Office, 2001, *Murang'a District 2001 KCSE Results: A Subject Analysis,* Murang'a: Murang'a District Education Office.

Mucai-Kattambo, V., Kabeberi-Macharia, J. and Kameri-Mbote, P., 1995, 'Law and the Status of Women in Kenya', in Janet Kabeberi-Macharia, ed., *Women, Laws and Practices in East Africa: Laying the Foundation.* Nairobi: Women and Law in East Africa.

Ndunda-Kivula, M., 2001, *Women's Agency and Education Policy: The Experiences of the Women of Kilome, Kenya,* New York: State University of New York Press.

Ndunda, M. and Munby, H., 1991, 'Because I Am a Woman: A Study of Culture, School and Futures in Science', in *Science Education,* Vol. 75, no. 6, pp. 683-699.

Odaga, A. and Heneveld, W. 1995, *Girls and Schools in sub-Saharan Africa: From Analyses to Action.* Washington DC: World Bank.

Okwemba, A., 2001, 'Parents Root for Single-Sex Schools', in *Daily Nation,* Nairobi, May 21, 2001.

Republic of Kenya, 1964, *Kenya Education Commission Report,* Nairobi: Government Printer.

Republic of Kenya, 1984, *8-4-4 System of Education,* Nairobi: Government Printer.

Republic of Kenya, 1988, *Sessional Paper No. 6 of 1988: Education and Manpower Training for the Next Decade and Beyond,* Nairobi: Government Printer.

Republic of Kenya, 1996, *Sessional Paper No. 2 of 1996: Industrial Transformation by the Year 2020,* Nairobi: Government Printer.

Republic of Kenya, 1998, *Master Plan on Education and Training: 1997-2010,* Nairobi: Jomo Kenyatta Foundation.

Riordan, C., 1985, 'Public and Catholic Schooling: The Effects of Gender Context Policy', in *American Journal of Education,* Vol. 93, no. 4, pp. 518-540.

Riordan, C., 1990, *Girls and Boys in School: Together or Separate*, New York: Columbia University Press.

Rosser, S. V., 1993, 'Female-Friendly Science: Including Women in Curricular Content and Pedagogy in Science', in *Journal of General Education,* Vol. 42, no. 3, pp. 191-220.

Schiefelbein, E. and Farrell, J., 1980, 'Women, Schooling and Work in Chile: Evidence from a Longitudinal Study', in *Comparative Education Review*, Vol. 24, no. 2/2, pp. 160-179.

Sifuna, D. N., 1990, *Development of Education in Africa: The Kenyan Experience*, Nairobi: Initiatives Publishers.

Sow, F., 1999, 'The Social Sciences in Africa and Gender Analysis', in A. Imam, A. Mama and F. Sow, eds. *Engendering African Social Sciences*, Dakar: CODESRIA, pp 31-79.

UNESCO, 2002, *The EFA Assessment: Kenya Country Report 2002*, Paris: UNESCO.

UNESCO, 2004, *The EFA Assessment: Kenya Country Rreport 2004*, Paris: UNESCO.

Wassanga, C., 1997, *The Attitude towards Science among Primary and Secondary School Students in Kenya*, Nairobi: Academy Science Publishers.

Wassanga, P. M., 2003, 'Female Participation and Performance in Mathematics, Science and Technology in Secondary and Tertiary Education in Kenya', *Proceedings of the 11th Annual Southern African Association for Research in Mathematics, Science and Technology Education*, pp.576-584.

Wong, K., Lam, Y. R. and Ho, L., 2002, 'The Effects of Schooling on Gender Differences', in *British Education Research Journal*, Vol. 28, no. 6, pp. 827-834.

9

Gendered Views of Science and Technology in the Performing Arts: Characterisation and Casting in the Kenya Schools Drama Festival Items

Lydia Ayako Mareri

Introduction

Kenya's education system has changed several times since independence. In the current education system, science is introduced to the students in upper primary school, and the students are exposed to scientific concepts ranging from physical to natural sciences. In secondary school, the students begin to study actual science subjects such as Chemistry, Physics and Biology. In their second or third year in secondary school, they make subject choices in readiness for the Kenya Certificate of Secondary Education (KCSE) examinations. Besides academic programmes, there are also co-curricular activities that the Ministry of Education regulates, including drama, music, athletics, ball games and various clubs and societies. These are available to all students. However, despite the availability of similar learning programmes and environments for girls and boys, girls do not perform as well as boys in science-oriented subjects as they progress from kindergarten, lower primary, upper primary school and on to secondary school. This is a global phenomenon and has had extensive coverage internationally. The explanations given have varied according to different research findings, but two main lines of enquiry have emerged: psychological and sociological (Kitetu 1998).

From observations made by teachers and parents alike, girls are usually interested in sciences when young. In kindergarten, they get involved in hands-on projects and draw pictures of themselves in an imagined future as doctors, archaeologists, marine biologists, etc. (Stueber 2002). But when the same girls are studied later in their schooling years, they are found to have lost interest in becoming scientists. The concern of researchers has been to finding out why this happens.

Girls' poor performance in the sciences at school results in women being under-represented in science and technology occupations (Eshiwani1983; Duncan 1989; Alele-Williams 1987) and in differential participation of the sexes in science education (Erinosho 1994). Appraising girls' science status has therefore taken centre focus in feminism. It has been argued that girls need to excel in science subjects so that they can be part of mainstream development and also because science leads to technological advancement, which is essential in human resource development (Erinosho 1994). Recommendations have therefore been made to involve the curriculum in bringing about the desired changes:

> Curriculum materials must be re-designed to ensure that they are relevant for boys and girls. At the same time, science teachers must be sensitized to treat girls and boys the same way in the classroom (Rathgeber 1995:185).

Here, the school has been identified as the vehicle for bringing about the desired change in the status of girls and women as a whole. However, although school systems are believed to be about cultural revolution and improving the social status of those who go through the systems, educational institutions have remained conservative and have shown show little willingness to depart from patriarchal societal set-ups (Kelly and Nilhen 1982). This is basically because teachers and students often bring into the school the patriarchal values, attitudes and beliefs learned from their social experiences at home and in the community as a whole.

This paper's main argument is that if some of the factors that hinder girls' access to science and technology are social, and there are social activities in schools such as drama festivals, then those activities can play a role in enabling girls to change their perceptions of science and technology, provided girls are given an opportunity to participate actively in the roles that express the changed perceptions. These perceptions can manifest themselves through characters created by the scriptwriters. Thus, there is a need to design and provide means through which the girls can be encouraged to develop an interest in science through what they already like. In view of this, a study was designed whose objectives were:

- to establish scriptwriters' awareness of the science and technological notions in the performing arts items
- to assess the characterisation and casting criteria used by scriptwriters and directors
- to determine the gendered tendencies in characterisation and casting for the scientific and technological roles.

The study was partly quantitative and partly qualitative. Quantitative data was suitable because certain information required actual numbers. However, qualitative data was suitable because the subjects of study were required to explain issues. So, this gave the respondents an opportunity to freely express their thoughts, attitudes and opinions. The responses were described and interpreted, and deductions were drawn. The expected responses included explanations of what was considered to be the greatest hindrance to providing equal opportunities to both girls and boys to take up science and technology roles in the drama items produced for the Kenya Schools drama festivals. The population included all the drama teachers in Kenya who are also scriptwriters and directors for the items usually presented by students during the annual schools drama festivals. A purposive sample from scriptwriters was obtained from the drama teachers attending the national workshop in August 2003. These represented all the eight provinces in Kenya. Two hundred drama trainers were invited, and 50 secondary school trainers participated in this study. The sample is summarized as follows:

- Female respondents were 9.5 percent, while 90.5 percent were male.

- The participants were from both single-sex and co-educational schools.

- They had all been involved in scripting for different periods as follows: 1 year and below—12 percent, 2-5 years—34 percent, 5 years and above—54 percent.

- The respondents had also participated in the festivals at various levels in different genres as summarised in Table 1 below.

Table 1: Percentage Distribution of Participation at Different Levels in the Four Genres

Level	Play	Dance	Verse	Narrative
Zonal	12%	6%	2%	2%
District	16%	4%	16%	10%
Provincial	24%	16%	22%	26%
National	24%	16%	26%	18%
Non-Participant	24%	58%	34%	44%

Table 1 shows that the highest participation was at the provincial and national levels. This is because the respondents were representing the eight provinces in Kenya. Table 1 also shows that there was high participation in the play category.

There were a total of 15 questions eliciting both quantitative and qualitative responses. The respondents were not expected to identify themselves, so their personal identities were not recorded. They were, however, both male and female, and all taught in secondary schools in Kenya and were involved in scripting for the schools drama festival. The types of schools in which they taught was not relevant for this study because the focus was on the conceptualisation of the ideas, characterisations and casting, regardless of whether the school was single-sex or co-educational. This was deliberately done in order to provide an opportunity for those who taught in single-sex schools to respond, guided by their own beliefs. Two instruments were used in this study, a short questionnaire from which quantitative data was obtained and a semi-structured interview from which qualitative data was obtained.

In the analysis of a character in drama, several things have to be considered: what the character does, what the character says, what other characters say about that character and, finally, the opinions of the scriptwriter and director. Thus, in order to analyse characterisation in the items, the scriptwriters were asked to state the kinds of characters they created for various roles. The analysis involved inferring meanings and implications from the narrations of the scriptwriters and the directors of the different genres of the drama items. This required them to indicate what influenced them when creating characters that expressed science and technology notions. They were also asked to explain how they determined who would take up the roles they created.

Social Influences on Girls' Interactions with Science and Technology

Socially, girls and boys are expected to behave differently. Children usually get explicit instructions on what is proper behaviour for girls and boys. Girls are told what is considered 'ladylike' and 'nice', while the boys are told what is expected of 'big strong men'. In fact, parents tend to punish aggression in their daughters and dependent or passive behaviour in boys (Weitzman 1982). When these gender expectations are transferred to classrooms, girls tend to depend on the boys to carry out experiments for them in science (Kitetu 1998). The gendered view of science is further perpetuated by science and technology being portrayed as a man's domain involving energy, dirty work and noise. Girls are not expected to be interested in such things (Lubega 1998). Thus, girls are implicitly discouraged from associating with science and technology activities, beginning with the classroom science subjects.

Besides being isolated from science subjects and experiments, teachers (male and female alike) tend to pay more attention to boys than girls in the classrooms. It has been established (Brophy and Good 1970) that the attention teachers give to pupils determines the development of self-confidence in pupils and contributes significantly to pupils' performance. Brophy and Good (1970) observed that boys have more interactions with the teacher than girls and appear generally salient in the teacher's perceptual field. These observations indicate that there is a discrepancy in the way teachers attend to girls and boys in the classroom. This isolation and stereotyping was even noticed as early as the 1960s (Weitzman 1982). School systems were observed as reinforcing sex-role stereotypes. For example, 73 percent of class readers were found to foreground male characters. When girls and women were included, they were represented as timid, inactive, unambitious and uncreative. They were also shown to be lazy twice as often as male counterparts and intellectually inferior. School textbooks were also found to retain these stereotypes (Weitzman and Rizzo 1974). Mathematics, science and even social studies textbooks purveyed an equally limited image of women. In science textbooks, only 6 percent of the pictures included adult women. This is how young girls in schools tended to be discouraged from science and channeled into more traditionally 'feminine' fields (Weitzman and Rizzo 1974).

This study went out of the classroom in order to establish the subtle factors that may contribute to the isolation of girls from interacting with science and technology. The findings have been used to establish the gender discrepancy in non-academic activities and how this discrepancy could determine the foundation laid for science and technology. The study identified performing arts (plays, dramatised dances, dramatised verses and narratives) as one of the key socialisation activities in the school setting. During the presentations of different items, there has been a notable use of technological and scientific notions and applications such as telephones, lighting techniques, sound effects, guns, computers, cars and medical equipment. Drama being arts based, it is expected that more girls than boys would take part, but it has been observed that there are usually more boys than girls participating in the items presented by co-educational schools in the drama festivals in Kenya. Moreover, in line with societal perceptions and expectations, boys are cast in roles that bear scientific and technological notions.

Community Theatre, Participatory Theatre or Theatre for Development

Does theatre matter in changing perceptions? Lately, there has been an emergence of different types of theatre in which the participants are expected to experience and learn various things that either educate them or enable them to acquire certain knowledge and skills. This has been referred to as theatre for development, participatory theatre or community theatre. Various theatre groups have been formed in different parts of the world and have used this type of theatre to address various social problems. Theatre for development generally aims at eliminating perceptions that militate against individual and communal change. It can be understood within the larger frameworks of participatory processes and consciousness transformation. In Kenya, theatre for development anticipates the de-conditioning and de-construction of oppressive conditions and situations that undermine individual and collective development (Desai 1991). For this kind of theatre to succeed, Byam (1999) argues that it must be framed within a philosophy and ideology that encourages change and empowers people to transform their lives. Over the years, a number of different approaches to theatre for development have been tried. Some of the major approaches are briefly discussed in the following sections.

Theatre of the Oppressed

Augustino Boal developed a new way of working with theatre, which he called 'theatre of the oppressed', during the 1950s and 1960s in Latin America (Epskamp 1989). He aimed to use theatre to empower oppressed people to change their situation by re-enacting their problems on stage. He explained that, instead of being spectators, they would become 'spect-actors'; instead of individuals allowing others to define their lives, they would make images of their lives themselves:

> To change the people from 'spectators'—passive beings in the theatrical phenomenon—into subjects, into actors, transformers of the dramatic action, ...the

liberated spectators, as a whole person, launches into action. No matter that the action is fictional, what matters is that it is action! (Boal 1979:122).

Oppression was defined as something that could be within you, as what Boal (1979) calls 'cops-in-the-head'. This internalised oppression is what prevents people from working together and changing their lives.

HIV/AIDS Communication

There are many theatre groups that have been involved in communicating HIV/ AIDS messages to communities in the third world. These include Amakhosi Theatre for Social change in Zimbabwe, Arepp Educational Trust in South Africa, Atelier-Theatre Burkinabe in Burkina Faso and many more. Theatre has proven to be an effective mode of reaching out to people with crucial messages on the dreaded disease. In Kenya, many organisations have used theatre activities to communicate HIV/AIDS messages to different communities. One such organisation is PATH-Kenya, which works with various theatre groups on outreach communication programmes, especially for the youth. PATH is funded by UNDP and has had successful outreach programmes with groups such as Artnet Waves in the Rift Valley and Kenya AIDS Intervention/Prevention Project Group (KAIPPG/International) in Western Kenya. The activities of these groups (and others) have attracted the attention of the youth and have enabled many young people to address HIV/AIDS issues more freely. HIV/AIDS theatre activities include plays, dances, narratives, poetry and puppetry, and they have been successful in communicating messages that official public health agencies have not been able to articulate. They have simplified the mysteries of HIV/AIDS and enabled those who participate to comprehend otherwise complex issues.

Human Rights Advocacy

There are many groups in Kenya that have used theatre activities to communicate human rights messages. The earliest known group was Kamiirithu, which operated under the umbrella of Kamiirithu Cultural and Educational Centre. The Kamiriithu Cultural and Educational Centre became a historical landmark in the history of theatre in Kenya and, specifically, theatre for development, when Ngugi wa Thiong'o and other facilitators from the University of Nairobi introduced the concept of theatre as a tool for development to peasants and workers in this impoverished village on the outskirts of Nairobi. As Ndigirigi (1999:71) notes, 'by having workers and peasants act in *Ngaahika Ndeenda*, the Kamiriithu group ... departed radically from the practice of other groups by having the underprivileged act in the drama about their lives'. Although the success of Kamiriithu has been attributed to its collective approach, it must be recognised that the presence of external facilitators with a vast knowledge of participatory education and theatre for development made a difference. The Kamiriithu project remains, however, a process of demystifying knowledge and bringing about realisation of reality.

Currently, the Kenya Human Rights Commission runs many theatre program-mes, including 'Human Rights Outreach', a project that uses popular theatre, including drama, puppetry, music, dance and poetry, as a tool of mass education on human rights, especially to generate public dialogue on human rights. The programme utili-ses theatre artists based in local communities and has proved very effective, although the participants have often found themselves involved in controversy due the political overtones of their activities.

Kenya National Schools and Colleges Drama Festival: Disguised Theatre for Development

The annual Kenya National Schools and Colleges Drama Festival is the single largest theatre event in East and Central Africa. It draws larger audiences to theatre spaces in several regions of the country than any other festival at any other time. The festival begins in February and continues to mid-April. The performances take place at zonal levels, then move through the districts to the provincial level and finally to the national level. The main performance genres include plays, dramatised verse, oral narratives and dramatised cultural dances. These performances deal with topical themes and issues that are of great concern to the society. Although the mode of presentation at the festival strictly follows traditional conventions of the proscenium arch theatre, it is interesting to note how much these theatrical pieces fit with the approach of theatre for development that Frank (1995) calls 'campaign theatre':

Campaign theatre (CT) is a form of theatre for development which is concerned with raising the consciousness of the people on such topics as childcare, environmental issues, health care, etc. The notion among the organisations which advance CT is that, with the help of theatre, a message will reach a larger number of people, and also that theatre, through its inherent entertainment value, is better suited to convey that message than, for instance, a series of lectures (78).

A major transformation of the Kenya National Schools and Colleges Drama Festival, which enhanced its capability as campaign theatre, occurred in 1981. A decision was taken to shift the national finals from the Kenya National Theatre in Nairobi, where it had been held for over twenty-one years, and rotate the site in the various provinces. This is how the festival still operates today. This meant that more people now have the chance to watch and learn from these highly educative and entertaining productions. From the foregoing, it is obvious that the festival does indeed perform some of the functions of theatre for development.

This study, therefore, aimed at finding out if the scriptwriters and directors of the items performed were aware that science, technology and the performing arts are complimentary and that none can exist without the other in the contexts in which they are studied and applied. The assumption was that, if performing arts have been used successfully to bring about change through theatre for development approaches, it should be possible to use the performing arts in educational settings to solve the problems that exist there. In Kenyan schools, both boys and girls take part in the drama festivals, and some of the roles students play articulate various scientific and technological applications. This means that those who take part in those roles have a chance to experience those applications as they participate. The results of the study are described, interpreted and discussed below.

Script Writers' Awareness of Science and Technological Notions

Responding to a question that required them to indicate if they were aware of the inclusion of scientific and technological ideas in plays, dances, verses and oral narratives, all the respondents indicated that they were aware. They also indicated that they utilised scientific and technological ideas by writing items that included technologies such as computers, industrial chemicals, bombs, radio and television, cameras, vehicles, medicines, lighting and sound effects, air travel and the use of telephones, etc. Ninety percent of the respondents indicated they were aware that these technologies were related to science subjects taught in schools. The majority singled out physics and chemistry as the basis of those ideas. In this regard, when they were asked if they consulted other subject teachers in the schools in preparation for the festivals and competitions, 96 percent indicated that they consulted, not at the scripting stage, but during rehearsals. Some of the subject teachers who were consulted included those in the subject areas that had been identified earlier. But they also said that they also consulted the language and art and design teachers.

Characters and Casting

The respondents were asked to indicate if, during the scripting process, they decided whether it would be boys or girls who took the roles of the main characters that contain the science or technological ideas that they scripted. Seventy-eight percent of the respondents indicated that, as they scripted, they considered specific individuals who would play the roles they were creating. However, 50 percent of the female respondents indicated that the female or male factor did not arise. Nevertheless, in the creation of those characters, both male and female respondents clearly indicated that they did not balance the male and female *characters* but created more male characters for those roles than female ones. Seventy-one percent specifically said that they created more male characters to take up those roles that articulate science and technological notions. The reason for creating more male characters was generally ascribed to female characters being 'weak' or 'difficult to deal with'.

Besides the creation of characters, the directors were asked if they ever encouraged girls to take up the roles that carried the scientific and technological notions such as computer applications, use of guns, medical careers and lighting systems. Seventy percent of the male and 60 percent of the female respondents indicated that they did not. Various reasons were given for not encouraging girls to take up these roles. The following are some of the actual responses to this question:

> It is hard to control girls or even a female character playing such roles. They make many mistakes and clumsily. Give a female character to use a gun and the results are disastrous …. She will hold it like a cob of maize and you can lose a lot of marks on character

credibility .... When you correct them repeatedly, they sulk and, as you know, we have a short period; I have no time to mother them (Respondent A).

You know as much as I do that most girls have short concentration span on these things. When a girl is playing a crucial role, you will die many deaths before you can have confidence in her .... The idea of working with a female for such roles is tedious, you need extra skills to succeed (Respondent B).

Boys are usually very cooperative and are obviously more confident in such operations. They don't usually worry about their bodies and such like things. They can also handle any emergencies without me getting ulcers during the festivals (Respondent C).

From such responses, we can see that the scriptwriters and directors of the drama items view girls as unreliable and unmanageable when playing science and technology roles.

When the respondents were asked to explain how they would ensure that girls also participated equally in the drama festivals, they fell back on the idea of genres, saying that there was a difference in the strengths of the various genres and that the girls would be more comfortable with dances and choral verses. The major reason brought out was that 'girls do well as a team', whereas the boys would do better in genres such as plays, narratives and solo verses because 'they are not easily distracted', 'they can handle emergencies well' and 'they are not as shy as girls'.

Applicability of Scientific and Technological Notions to Science Subjects

While preparing for the performances, the respondents were asked to indicate if they encouraged the performers to relate the ideas they were articulating on the stage to real-life situations and subjects they learned in class. Sixty-five percent said they did so and also said they helped them understand that those subjects would enable them to go into careers like medicine, electricity, woodwork and computers and information technology. Generally, they indicated that they encouraged them indirectly by saying such things as 'you are doing well', 'bring out that section like the engineer/doctor you are' or 'do that like you were in the examination room'. These are encouraging utterances and would make the performers reflect on the parts that they play and infer their importance.

Asked to indicate how their performers did in both science and arts subjects, they responded in a variety of ways. Most said their performers generally enjoyed arts subjects more than science subjects and did better in them. The measure of 'enjoy' was, however, gauged against class performance. A few of the respondents indicated that some of the best science students were their main characters in the items that they performed during the Kenya National Drama Festivals. This, they indicated, was because they interpreted the science concepts with ease.

Can Participation in School Drama Festivals Enhance Performance in Science Subjects?

The respondents were asked to suggest ways in which participation in school drama festivals could help enhance students' performance in academic science subjects. Table 2 summarises the responses. The responses were placed in five clusters as shown in the table. The clusters refer to the possible ways in which the students can experience science notions through drama. The responses were varied, so the clusters imply they referred to the terms included in Table 2.

Table 2: Clusters of Comments on How Participation in Drama Can Enhance Performance in Science Subjects

Cluster	Percentage
Themes and theories	21%
Props, sound effectsand stage craft	24%
Participation (attitude and confidence)	53%
Other	2%

A slight majority (53 percent) of the respondents indicated that participation of students in drama items could help them improve their attitudes and gain confidence when dealing with science and technology subjects in classroom situations. They indicated that this would happen when the students transferred their creativity in drama to activities in science and technology. Table 2 also shows that 24 percent of the respondents indicated that students could relate the themes selected and the theories of drama applied in their festival performances to classroom science and technology subjects. Twenty-one percent indicated that using the props in drama would be helpful in understanding the practical aspects of science and technology concepts. Two percent of the responses did not fit in any of the five clusters. These include comments about science being beauty and about the value of creating scenery for a production, which should be beautiful because it uses scientific inventions.

From Table 2, therefore, we can observe that a majority of the scriptwriters recognised that there is a relationship between the performing arts and science and technology applications. This observation can be solidified by quoting some of the actual responses as follows:

- Using science-oriented items can help the students create their own relevant items. When given roles of science ideas, they have to consult the people with science knowledge (Respondent D).

- Use of sound tracks and lighting exposes them to the beauty that can be created by science (Respondent E).

- They will change their attitude or perception of science by playing the science roles (Respondent F).

- By scripting plays with scientists, like Darwin, as characters, the main ideas involve evolution (Respondent G).

Discussion

The findings above indicate that there are more male drama scriptwriters and directors than female ones. Although it is an arts-based activity, male scriptwriters and directors dominate it. This could be because of the implied scientific and technological applications, a possibility that can be inferred from the fact that all the scriptwriters, both males and females, were aware of the scientific and technological applications of the material they included in their scripts. They knew that those applications had science bases and that they had a relationship with science subjects. They also indicated that they consulted science teachers for their productions. So the scriptwriters know that their items have scientific and technological ideas which the students are expected to articulate during the drama festival presentations. This leads us to infer that the scriptwriters tend to create science and technology ideas specifically for male performers to articulate, since they create more male characters during scripting in readiness for male participation. This implies that fewer girls will get the opportunity to take up roles that articulate scientific and technological concepts, because the scriptwriters prefer male performers for those roles.

The majority of scriptwriters, who were also the directors of the items, tended to create roles that articulate scientific and technological ideas and assign them to preselected individual performers according to their assumed theatrical abilities. Boys and girls are not given an equal opportunity to compete for those roles. They reinforce this practice by creating more male characters than female ones. This means that, even if the script were written for a girls' school, there would be more male roles. This is why assigning of roles is sometimes done by typecasting, as indicated by some directors in such responses as 'the characters that I want' or 'obvious ones'.

From the results presented above, we can also infer that majority of the scriptwriters are aware they are required to encourage the artists to relate to real-life situations and the science subjects studied in school. This means that they are also aware that the ideas they create have a direct relationship with science subjects and that these ideas could easily be applied to technologies like communication systems, carpentry, woodwork, metalwork and also computer operations. The scriptwriters were also aware that the roles the performers play are directly linked to scientific and technological careers. This is why they encouraged the artists who took part to identify themselves with the scientific professions that were implied in the roles that they played like doctors and engineers.

It is also implied that the majority of scriptwriters and directors of the school drama festival items do not encourage girls to take up roles in which scientific and technological concepts are embedded. This means that they do not think that girls are capable of conceptualising and articulating those scientifically and technologically oriented ideas effectively, suggesting that the scriptwriters and directors believe that the girls are weaker than the boys and cannot be trusted to articulate the scientific ideas effectively. This is demonstrated by one of the responses, which indicated that 'I will keep my fingers crossed'.

From the results presented, we can also infer that the scriptwriters and directors of the school drama festival items know that participating in the items provides an opportunity for the performers to experience the ideas they communicate to the audience. The information becomes part of them. This means that those who take part get a chance to internalise the ideas. Consequently, when they participate in the items that contain science and technology ideas, they experience and absorb those ideas. Thus, those who take part in those items that articulate science and technology ideas are likely to build confidence about, and cultivate desirable attitudes towards, the subjects related to the ideas that they articulate on stage. Their confidence will even be stronger if the items that they take part in win awards at the festivals. This will encourage them to experiment and try to apply what they have experienced. This is demonstrated by the observation of one respondent that 'they can easily do various experiments on their own and improvise props by applying their knowledge in the science subjects' and the comment by another that 'they will change their attitude or perception of science by playing the science roles'. However, despite the scriptwriters' knowledge about the usefulness of participating in items that contain scientific and technological notions, they still isolate girls. This denies girls an equal chance to participate in those items and experience the ideas therein.

Conclusion

We observed earlier in this paper that schools have failed to meet their expectation of being able to provide conditions through which gender stereotypes can be eradicated. It has also been observed that schools in Kenya provide social settings through which students at secondary level interact. There are many social activities in the school setting. This paper focused on the drama festival and the items that the students participate in, which are usually scripted and directed by their teachers. We have seen from the findings that the schools have not been able to provide friendly environments with equal opportunities for girls and the boys to access science and technology concepts even in the so-called social activities within the school. The scriptwriters share those stereotypes about girls being weak and not being able to measure up to the required standards. They will therefore not let them handle critical roles in the drama items for the festivals.

However, the scriptwriters also know that drama can enable the students to interact with science and technology subjects more easily, if they are exposed to similar concepts in drama items. Despite that knowledge, the teachers as scriptwriters seem to be perpetuating the isolation of female students from science and technology activities. This is done through scripting science concepts to be played by male rather than female characters. In reference to the literature cited, drama activities have been used to enable individuals to change their attitudes and perceptions. The scriptwriters and directors of drama items for the schools drama festivals indicated that participation can change the perceptions and attitudes of those who participate. In the same way, we can conclude that girls and boys can be given equal opportunities to participate in the science and technology drama items. This will enable them to experience the concepts equally and in turn encourage them to develop more positive attitudes towards related science and technology subjects. We have also seen that the drama teachers tend to isolate girls from interacting with the science and technology ideas during the drama festival activities. This is consistent with the earlier observation that teachers pay little or no attention to female learners. This situation can change if the teachers are made aware of the consequences of their actions.

The findings of this research suggests that out-of-class activities and students' participation in these should be scrutinised to see how, if at all, they challenge and are helping change societal gendered perceptions on science and technology. More specifically, in relation to the results and the conclusions of the study presented above, the following recommendations can be made:

1. The Kenya Schools Drama Festival scriptwriters need to be sensitised about the usefulness of balancing the male and female characters when they are scripting, so that they provide equal opportunities to both boys and girls to experience the science and technology ideas they include in their scripts.

2. The directors of the items that are presented at the festival should be encouraged to audition both girls and boys for the main roles that they create in order to give equal opportunities to the performers regardless of their sex.

3. The scriptwriters and directors should not ignore girls or discourage them from taking up roles in which scientific and technological concepts are embedded. Instead, girls should be encouraged, if they show interest.

4. The writers should also script without any gender discrimination. This will provide more opportunities for girls to experience the scientific ideas and enable them to enhance their interaction with science and technology subjects.

5. A special award should be created in all categories and genres for items that articulate scientific ideas through balanced gender characterisation.

6. The Kenya National Drama Festival Association needs to restate the objectives of participating in the festival so that they could include enabling the students to develop desirable attitudes towards academic subjects, with specific reference to science and technology.

7. Further research should be undertaken in order to find out what the girls and boys themselves have to say about their experiences, especially in items that reflect scientific and technological ideas. Further research should also be undertaken to study the finished scripts so as to analyse the discourse used and compare it with the findings in this study.

References

Alele-Williams, G., 1987, 'Science, Technology and Mathematics Education for All, including Women in Africa', keynote address at the Commonwealth Regional Workshop on Women and STM Education, Accra.

Boal, A., 1979, *Theatre of the Oppressed*, London: Pluto Press.

Brophy, J. E. and Good, T. L., 1970, 'Teachers' Communications of Differential Expectations for Children's Classroom Performance: Some Behavioural Data', in *Journal of Educational Psychology,* Vol. 61, pp. 365-74.

Byam, L. D., 1999, *Community in Motion: Theatre for Development in Africa,* London: Bergin and Garvey.

Desai, G., 1991, 'Theatre Development in Africa', in *Ral Fall,* Vol. 22, no. 3, pp. 7-9.

Duncan, W., 1989, *Engendering School Learning: Science, Attitudes and Achievement among Girls and Boys in Botswana,* (Studies in Comparative and International Education 16), Stockholm: University of Stockholm.

Epskamp, P. K., 1989, *Theatre in Search of Social Change,* The Hague: CESO.

Erinosho, S. Y., 1994, *Girls and Science Education in Nigeria,* Abeokuta: Ango International Publishing.

Frank, M., 1995, *AIDS Education through Theatre: Case Studies from Uganda,* Bayreuth African Studies, 35, Bayreuth: Bayreuth University.

Kelly, G. and Nilhen, A., 1982, 'Schooling and the Reproduction of Patriarchy: Unequal Work, Unequal Rewards', in Apple, M., ed., *Cultural and Economic Reproduction Education,* London: Routledge.

Kitetu, C., 1998, 'An Examination of Physics Classroom Discourse Practice and the Construction of Gendered Identities in a Kenyan Secondary School', unpublished PhD. thesis, Lancaster University.

Lubega, M. K., 1998), 'Gender Bias in Education: Challenges to Teachers and Teacher Educators', in *Teacher Educator,* Vol. 11, pp. 11-19.

Ndigirigi, G., 1999, 'Kenya Theatre after Kamiirithu' in *The Drama Review,* Vol. 43, no. 2, pp. 72-93.

Rathgeber, E. M., 1995, 'Schooling for What? Education and Career for Women in Science, Technology and Engineering', in G. Oldham, ed., *Missing Links: Gender Equity in Science and Technology for Development,* Ottawa: International Development Research Centre.

Stueber, N., 2002, 'Testimony for the Hearing on Women in Science and Technology', speech delivered to the Subcommittee on Science, Technology and Space of the Senate Committee on Commerce, Science and Transportation, Portland, OR.

Weitzman, L. J., 1982, 'Sex Role Socialisation', in J. Muff, *Socialisation, Sexism, and Stereotyping,* Washington D.C: C.V. Mosby.

Weitzman, L. J. and Rizzo, D., 1974, *Images of Males and Females in Elementary School Textbooks,* New York: National Organisation for Women Legal Defence Fund.

10

Repositioning Computer Studies: Cultural Context and Gendered Subject Choices in Kenya

Fibian Kavulani Lukalo

Introduction

> If Marie Antoinette haughtily advised the French authorities in the late 18th century to let the poor eat cake when they asked for bread ... the rich nations in the early 21st century shout, 'let them have computers', when the poor of the world ask for food and respect (Nettleford 2002:25).

Education is gendered, and the developed West often influences educational systems in Africa. Therefore, the sentiments expressed by Rex Nettleford, Vice-Chancellor of the University of the West Indies, point to the various socio-economic and cultural paradigms that ameliorate the advancement of computer technology in Africa. Nettleford's argument, when assessed within current global educational debates, reflects the continued pernicious effect of computer technology and its centrality in the advancement of science and technology, which continue to be unattended to in Africa. Additionally, for all intents and purposes, gender, the workings of agency and structure in education, is often not theorised into this science and technology debate in terms of access and practical needs.

The need for enhanced Science and Technology (S&T) in Kenya led to the introduction of computer studies in Kenyan schools in 1994. This move was primarily concerned with teaching computers as an optional science-oriented subject. The content of computer studies covers both basic computer literacy and computer engineering at higher levels. Inevitably, its categorisation as a science subject, and the lack of understanding of its implementation, merely perpetuates constraints already imposed on girls' participation by social structures. When mapped against the educational backdrop of gender-oriented biases, girls' exclusion and disparities in terms of geographical regions, school resources, available infrastructure (electricity, water, laboratories, land space) and teaching personnel, the inclusion of computer studies was not without obstacles. Therefore, it came as no surprise that the first national examinations to include computer studies took place in 1998 with only twenty-two students registered. Such hesitation in the practical implementation of computer studies raises critical questions related to the educational system, subject choices, subject clusters and gender.

This article arises out of a concern that the cultural, social and economic context in which educational policy with regard to computer studies has been developed, together with the manner in which the subject has been introduced, learned and practiced, contributes to the perpetuation of gendered disadvantage in S&T in Kenya. One interpretation may be that the teaching of computer studies has been left to those schools and geographical regions where appropriate infrastructure is available, in particular urban schools. Another interpretation might be that the subject highly augments power structures in relation to types of schools (private, public, district, provincial or national), gender and external linkages. These observations depict an educational system that is favourable towards advantaged schools and facing difficulties in promoting a truly gendered approach.

The need to examine educational policy with respect to S&T, especially computer studies, coupled with the ways in which these policies reflect gender disparities and gender role expectations in the wider society, therefore adversely affecting the participation of women and girls in computer studies in Kenya, is the focus of this chapter. Focusing on selected Kenyan secondary school learners in Forms One–Four in Western Province, the chapter seeks to articulate and discuss the relationships and challenges between gender and S&T that appear to emanate from a variety of contexts. The data collected includes educational consultant interviews with two members of the Kenya National Examinations Council (KNEC), along with both informal and structured interviews with students and staff at four selected secondary schools. Published aggregate computer studies statistical results from the 1998–2002 Kenya Certificate of Secondary Examinations (KCSE) were obtained from the KNEC.

Macro-structural features, that is, type of school, academic performance, gender and subjects offered, presented a further basis for evaluating the ways in which gender is experienced. Type of school refers to the secondary school categorisation adopted by KNEC, that is, private schools, district schools (76 districts), provincial schools (8 provinces) and the seventeen national schools. Whereas the national schools select the top students nationwide, the district and provincial schools must select 85 percent of their students from their own district or provincial catchment areas.

A questionnaire was developed to obtain preliminary knowledge of girls' computer studies expectations. Using an identified set of contextual factors that appear to ameliorate girls' entry into S&T (Kakonge 2000), the general objectives of the computer studies subject were reflected upon as a way of problematising the concept of gender. Ultimately, a framework was developed for understanding the gendered tendencies embedded in educational policy on computer studies and the contextual factors that continue to promote the exclusion of girls in S&T in Kenya.

Reconstructing the Science and Technology Debate

While every society has a wealth of information and a rich knowledge-base of its own, the power to own, structure, decide upon and control the technology for coding, transmitting and using information is determined by economic means, with the dividing line following economic, linguistic, ethnic and gender barriers (UNESCO 2003:68).

A considerable amount of research point to gender differences in educational policies and their implementation (Harding and McGregor 1995; Lukalo 1999). In Kenya, Eshiwani showed in the 1980s that the underlying determinant of women's under-representation in S&T disciplines was the inadequate preparation provided at the secondary school level (Kakonge 2000). Various studies have indicated that access to education in Africa has seen girls marginalised and generally under-represented in the fields of S&T in secondary schools and tertiary institutions (Jegede et al. 1996; Truscot 1994; Makau 1999). Reflecting on various African countries, Mariro (1999) found the scientific subject option characteristic in S&T-related subjects in secondary schools a critical catalyst in girls' further isolation from S&T careers. In addition to these findings, educational advancement in science subjects at the KCSE level in Kenya testify to social factors, such as household chores, school drop-out rates, early marriages, length of time taken to qualify for S&T careers, that are related to girls' adult family roles and continue to inhibit their progressive learning. Conversely, these factors have significant effects on performance and educational outcomes for female learners but not for male (Njenga 1999:164).

Other issues that give credence to the view that gender-related factors continue to interfere with girls' representation in S&T fields include lack of positive female role models, inadequate science facilities, socio-cultural expectations, gendered stereotypes and gender expectations by teachers, parents and the community (Erinosho 1994). To interrogate the very nature of the notions of S&T depends on the cultural and social expectations of the gendered roles for girls and the existing masculine conception of the scientific enterprise, which deters girls from pursuing S&T choices. Sands (1993:238–248) and Malcolm (1993:249–253) argue from the premise that the manner in which gender is experienced, fought over, sought after, worked at, accomplished, endeavoured and negotiated at a micro-level in girls' day-to-day lives is critical in contesting assumptions of gender and S&T.

Experiences in S&T for learners at the secondary school level are both meaningful and important, since they have significant consequences for their career outcomes. These outcomes can be explained in Harding's (1986) formulations of construction of gender as structural and symbolic. The structural is manifested by the gendered division of labour (arts subjects — female, S&T subjects — male), while the symbolic manifests itself through the languages and images that surround S&T (Malcolm 1993; Kitetu 2003). This gender division is culturally specific, socially constructed, linked to gender ideology and interwoven into broader social processes and power relations (Stepulevage 2001). Therefore, the question of how gender influences S&T subject choices in secondary schools in Kenya calls for a critical analysis of the constraints imposed on girls' participation by social structures, hegemonic social identities and the distribution of educational privileges.

As Enos (1995) argues, the structural adjustment programmes levelled at Africa in the 1980s have subtly dictated the choices of areas to be pursued in S&T based on the benefits for the developed countries. Ultimately, the indirect consequences of these policies filter down to the 'grassroots' level and lead to the exclusion of women (Beoku-Betts 2003). The feminist theme that 'science is socially constructed and as such embodies a history and a political agenda' rings true (Olson 1994:78). The field of education has been a locus for counter-practices and counter-discourses that have perpetuated gender exclusion tendencies (Obanya 1999). Researchers often look for explanations for the gendered exclusions in education in different ways with the aim of deconstructing upheld notions and 'truths' (Makhubu 1993). A key assumption of these approaches in post-colonial educational studies in Africa is that culture is constituted by the learned behaviour passed down from generation to generation. Such power to participate effectively in any process of negotiation has historically been divided along gendered lines.

Given the rate at which many research bodies were set up for African women (e.g., ACTRW, under the auspices of the UN's Economic Commission for Africa, in 1995) or by the African Woman (e.g., AAWORD 1977; FAWE 1999), gender continues to pose theoretical, analytical and practical challenges to Africa. Third-World women's activities tend to get devalued and depoliticised when examined through the binoculars of Western feminists. As a result, hegemonic Western feminist discourses end up framing Third-World feminist issues. Citing the heterogeneous nature of women's lives in developing countries, Mohanty discusses women's lives as a source of both compliance with and resistance to the dominant relations of power. When girls are offered subject choices from a fettered position (school setting, variety of subjects available), factors such as low self-esteem and low motivation enhance the gendered discriminatory tendencies. Thus, understanding the structured realities as they exist at local and national levels is critical.

Positioning the Computer Debate

When the teaching of S&T is presented to learners as a male domain, the social constructions of specific subjects in the classroom take on constructions of gender (Makhubu 1993). These gendered constructions have been aptly discussed by Grint and Gill (1995), Beoku-Betts (2003), Galpin (1992), and Kitetu (2003). These and other studies of the role of S&T in education have been diverse and have adopted varies theoretical underpinnings. What is clear is that, as Harding (1991:50) argues, scientific and technological changes 'are inherently political, since they redistribute costs and benefits of access to ... resources in new ways. They tend to widen any pre-existing gaps between the haves and the have-nots unless issues of just distribution are directly addressed'. Significantly, unlike other areas of educational technology change, the introduction of computers into schools was largely uncontested. A utopian wave of enthusiasm presupposed that computers were inherently and unequivocally a 'good thing' for education in Kenya and a sure prescription for industrialisation by 2020 (Coughlin and Ikaria 1988). However, this view ignores questions of the social, cultural and historical construction of information technology and 'the social meanings of science and technology [rooted in] ... the curious coincidence between masculinism and Eurocentrism' (Harding 1991:245). Ignored also is the issue of life outside the formal education system, where everyday life involves learning, creating and negotiating with technology (Omoka 1991).

The S&T debate in Kenya cannot afford to decontextualise computing from the wider social and political variables that shape the gendered contexts of human resources and the school itself (Beoku-Betts 2003). As White, Shade and Brayton (2001:50) point out, 'for women and men alike, computers exist in an uneasy societal space ... [where] fear and anxiety ... [are] connected to a lack of personal control over the impact computers have on our lives'. The feeling of lack of control for learners in Africa accentuates the dependency these learners have towards the Western world of computer innovations. In this respect, many women experience disparities in relation to computer access, ICT distribution and technology (Wakhungu 2003). White, Shade and Brayton (2001) highlight ongoing concerns about computer use in developing Nations creating access barriers, together with the corresponding need to enhance, rather than detract from, indigenous knowledge. Green and Adam (2001) show how technology has favoured masculine practices and has treated feminine practices as subterranean. For S&T in Africa, the absence of vital data about gender and technology relations enhances continued practice of unfriendly policies that deter women's participation. Grint and Gill (1995) suggest that, even in the West, technology has ignored how feminine practices utilise technology in different ways. In schools, phenomena such as the 'Pygmalion Syndrome' and other experiences influence one's decision to choose or not to choose computer studies.

Gender, Culture and Location: Intersections and Relations to Computer Studies

Computer studies in the formal education system has a distinct presence, but the process of engagement with the computer is continuous and interwoven in everyday life. In reflecting on this subject, the identification of spaces for agency that inform S&T debate is crucial. Diakitie (1991:123-127) cites the example of ICT policies as a major drawback to the development of technology in Africa, noting the long history of S&T subjects being used as agents of social control, especially of the poor. What about computer studies in Kenya since its inception at the secondary school level in 1995? Since the secondary school level takes four years of learning, the results first began to emerge only in 1998 and, as Table 1 below shows, female candidature was higher than that of male.

Table 1: KCSE Computer Studies Performance (1998)

Candidature			Type of School			
Female	Male	Total	Pri	Dis.	Pro.	Nat
17	05	22	-	-	2	-

Note: Pri-Private, Dis-District, Pro-Provincial, Nat-National
Source: Kenya National Examinations Council

As Table 1 shows, all the candidates came from only two schools, both belonging to the 'privileged' provincial category of schools. A visit to these two pioneering schools revealed that many of the computers in use were donations from religious organisations in America. The schools had to overcome technological obstacles such as outdated hardware and software, lack of backup facilities and even a lack of technicians. Most of the 17 female candidates saw their choice of the subject as a marker of the changing times and a head start to secretarial careers. Thus, the computer donations, despite their myriad difficulties of installation and maintenance, acted as agency in relation to constructing familiarity with technological knowledge and skills in these pioneering schools.

Gladys Nasongo, a KNEC officer, argued that the higher female candidature in 1998 testified to the encroachment of women in Kenya on S&T, thus ultimately aiming at devaluing masculinist tendencies in the area. The message sent by the results in 1998, she said, was that computer studies was a feminine domain, thus appropriating a female identity tag. When this gendered identity notion infiltrated the everyday educational structural relations and cultural meaning system, computer Studies became the symbolic agent of deconstructed masculine technology. It is this feminine engagement with computer studies technology that became the 'origin story' (Haraway 1991). However, from another perspective these 'privileged' provincial schools aided in constructing young women's opportunities away from the male-dominated fields of S&T into the 'softer' ground of computer studies (Makhubu 1993).

It is interesting to note that the seventeen national schools, despite all their resources, were slow in offering the computer science option, only registering candidates in 2000. A closer examination of the candidates in these schools reveals lower female candidature by 2002, 91(30 percent) female candidates to 213 (70 percent) male candidates. National secondary schools in Kenya belong to highly privileged league of schools, many relying on the goodwill of the government and well-endowed PTAs. Whereas other secondary schools in Kenya can afford to offer only nine subjects at the KCSE level, National schools offer at least eighteen subjects. This translates into smaller classes, wider subject choices, lower teacher-pupil ratios, more varied appeal to all types of learners and better learning facilities. Thus, when fewer girls enrol for computer studies in national schools, they end up 'self-selecting' out of S&T careers. Paradoxically, this suggests that other institutional barriers away from school act as deterrents in the S&T choices of these girls.

Several dimensions of social change seem to have impacted on the numbers of candidates for the KCSE examinations from 1998-2002 (see Table 2). First, the proliferation of 'computer technology strategic propaganda' nationally and globally promoted enrolment, which registered a tremendous increase from 22 candidates in 1998 to 2,145 candidates by 2002. However, while more students enrolled for computer studies in private, provincial and national schools, the district schools category, mostly found in rural areas, experienced low enrolments for various reasons, lack of electricity, infrastructure and computer hardware/software, along with low levels of economic assistance from government and PTAs.

Table 2: KCSE Candidate Enrolment by School Type (1998–2002)

School Type	Years/Candidature										
	1999		2000		2001		2002		1999–2002		
	F	M	F	M	F	M	F	M	F	M	Total
Pri	15	09	72	103	106	149	237	302	430	563	993
Dis	17	07	57	07	42	69	109	94	225	177	402
Pro	19	35	153	132	325	273	690	409	1187	849	2036
Nat	-	-	32	43	55	61	91	213	178	317	495
Total	51	51	314	285	528	552	1127	1018	2020	1906	3926
(%)	50	50	52	48	49	51	53	47	51	49	
Grand Total	102		599		1080		2145		3926		

Note: F-Female, Pri-Private, Pro-Provincial, M-Male, Dis-District, Nat-National.
Source: Kenya National Examinations Council.

One headmistress from a provincial girls' school had this to say about the resources required for effective computer studies programmes:

- The parents in my school have been very helpful in funding the computer project. We started off in 1997 with two computers, and since then have added to the present number of forty. Without the computers and a good teacher, this subject is not easy.

The headmistress' perspective demonstrates the imbalances that exist in the facilitation of computer studies. As her comment makes clear, other factors than gender hamper the implementation of the subject. Thus, the relationship between the community's resources and aspirations is critical for computer science. Clearly, the social outlook of the community helps shape the expectations of a school's subject choice array. Judging from the high level of female enrolment by 2002, one can conclude that the wider Kenyan community is receptive to the need to facilitate more girls in S&T-

related domains. Commenting on the astronomical increase of students in computer studies, a KNEC official had this to say:

> Many schools currently offering computer studies have received a lot of donations in terms of the hardware and software from organisations abroad. Through established connections, either church-based, NGOs or individuals, outdated computers have found their way into our schools. In other instances, through school PTAs, certain levies have been forced onto parents to enable schools to buy computers. Whereas you can control what you buy, you can never control a donated piece of equipment.

Importantly, the KNEC official draws back the S&T debate to what Omoka (1991) describes as the political dynamics and dependency of technology transfer. Increasingly, PTAs, community groups, NGOs and philanthropic organisations are bridging the economic gap and hegemonic social space that the Ministry of Education should fight in the provision of equipment to schools. The process creates new links between schools and these agencies and has implications for new thinking about agency because of the ways in which the agents utilise the power they possess to change communities. The assumed value of computer technology acts as a positive facilitator of change. However, in other schools, the belief that technology is expensive acts as a barrier. In this respect, one particularly interesting aspect of computer studies in schools is the enrolment by gender since 1998 (see Table 3).

Table 3: KCSE Gender Enrolment Pattern 1998–2002

Year	School Type			
	Female	Male	Co-ed	Total
1998	1-50%	1-50%	0	2
1999	4-33%	4-33%	4-33%	12
2000	11-26%	15-37%	15-37%	41
2001	29-40%	27-37%	17-23%	73
2002	57-39%	52-36%	37-25%	146
Total	102-37%	99-36%	73-27%	274

Source: Kenya National Examinations Council

As Table 3 indicates, there are more female schools offering computer studies at KCSE level in Kenya, and this serves as an entry point for girls to increasingly tap into the field of 'soft' technology. Thus, the male aura of S&T education, which acts as one obstacle to girls' attainment of computer literacy, does not have to be accepted as given. One of the main analytical points in this respect is that agents (communities, NGOs, donors) act out of the gendered social and individual transformations of their views of technology. It can be noted that, quantitatively (in terms of schools, student numbers, education offered) more girls in the 1998-2002 period were educated in district (225, or 56 percent girls/177, or 44 percent boys) and provincial (1187, or 58 percent girls/849, or 42 percent boys) schools.

Qualitatively, in terms of computer studies teachers, facilities infrastructure and financial status, the majority of these schools compare unfavourably to private schools (563 boys/430 girls), and national schools (317 boys/178 girls). This structured categorisation of schools, considering also that a substantial number of secondary schools offering computer studies are co-educational, increases the marginalisation of girls in an educational system of inequality. The influence of location is pivotal, since national and private schools enjoy political, social and economic power in an area of study structurally dominated by men. In these schools, the computer 'symbolically' becomes a male enterprise.

Adopting Hardings' (1986) multi-level theory of gender, we further took into account gender processes at the structural, symbolic and identity levels by mapping the gendered nature of the computer studies teaching force and the experiences of the head teachers of the sampled secondary schools. Gender, symbolically epitomised by the teaching force, can enhance masculine/feminine values and connotations. In 2002, of the thirty-three secondary schools in Western Province offering computer studies, there were only two female computer studies teachers in district girls' schools. The situation was similar in other provinces, clearly showing a field competitively tapped into by female students yet dominated by a male teaching force. The two female teachers, when interviewed, gave their views as follows:

> **Female Teacher A**: I developed interest in this subject at the university and worked on my own. My girls appreciate what they are doing, but the majority just think about secretarial jobs. I have had to tell them that other possibilities exist. Most of the girls had never touched a computer before, so I am happy their awareness level has increased. It is these girls I use to tell others what computer studies means.

> **Female Teacher B**: It has a lot to do with the school. The computers are not enough, so we regulate the number of girls taking the subject. Also, only the bright and high achievers are selected for this subject. I have to keep reading and know what is on the market for my girls. Often. I have to take them to visit other schools.

The choice of studying computers is strongly linked to user values that may have been constructed consciously or unconsciously. Choices made by students can be influenced from outside the formal school system and may find grounding in everyday life experiences. Teacher A's comments above bring out the fallacy of women's technological knowledge being conceptualised essentially as women's productive activity, and this needs to be interrogated. Teacher B identifies the school and its procedures for selecting learners (high achievers) as important to the study of computer science. Such macro-structural features intersect with the cultural models of S&T and frame the academic decision-making process for the learner. Ultimately, this helps perpetuate a male-dominated S&T community, as was affirmed by two teachers in a national boys' school:

> **Male Teacher A**: Some of the students already have exposure to computers before joining secondary school. Another percentage wants to enroll for a Diploma in Computer Science after school. This subject, being an option and not necessarily key to admission into science degree programmes at the university, most of our students

prefer the physics option, because it is really the bedrock for joining science-based professions.

From this teachers' comments, one can deduce that, even before joining Form One, some boys already identify with computers. Others project their vision towards getting tangible certification of their skills. Comparatively, the physics option is not a popular science option among girls in secondary schools (KNEC: 2002). As a whole, it can be understood as part of the symbolic dimensions of that feed into gender identity. The teacher aptly points out that individual choice dictates what some students want to pursue, augmenting the view that the S&T gender gap is the outcome of internal, individual processes and academic choices. The schools and educational policies recontextualise the gender socialisation processes in terms of gender-appropriate subject matter for out-of-school S&T career success.

The total female KCSE enrolment for computer studies from 1998 to 2002 (52 percent female/48 percent male) makes it clear that computer studies appeals to more female candidates than biology and chemistry. This can be construed to mean that higher female enrolment in computer studies is an example of the hidden gendered assumptions that organise educational practices. Gender inequalities, specifically the gendered cultural and socialisation process in the home, are the edifice upon which educational systems are built in Kenya. Schools depend on the parents, and particularly mothers, to be involved in the decision-making process of subject choices for their children. Some of these choices sustain or advance gender inequities. When the number of female schools is compared to the mixed schools (1999–2002), the following is observed:

1999 – 4 girls' schools, 4 mixed schools,

2000 – 11 girls' schools, 15 mixed schools

2001 – 29 girls' schools, 17 mixed schools

2002 – 57 girls' schools, 37 mixed schools

These figures portray the choice of Computer Studies as a 'softer option' for female students consistent with the view of gender ideologies about women's role as 'caretakers' in S&T fields. The figures invite one to interrogate this view in two related ways. First, the activity of subject a 'choice' at KCSE level is indistinguishable from the culturally perceived role of women. The classroom nuances of mixed schools, and their discriminatory practices in classroom discourse, teacher attention (Kitetu 2003) and societal expectations for women influence these choices. The implied connection between the internal world of the classroom/school and the world of external expectations is notable.

Essentialising Individual Aspirations

Several learners were interviewed on the question of computer science subject choice and the general objectives of the subject. The following discussion presents and analyses some of their comments.

Vuyanzi (private mixed school): Everybody knows that without computer skills in today's world, one is doomed! I want to be able to use my computer skills, when I leave for further studies to England next year …. I want to study financial banking and, as you see, the computer is dictating many things …. I need this subject, that's me!

Vuyanzi does not take on board her lived experiences of gender difference, and perhaps her views can be seen within Kitetu's (2003) 'privileged femininity' perspective. Vuyanzi sees herself benefiting from the recent S&T changes in terms of opportunities available to all, not just women. She views computer studies as integral to the sense of purpose that accompanies her dreams of upward mobility. Thus, her choice of computer studies is meaningful and important to her because computer studies entails the potential to have a significant consequence on her career outcomes. By affirming 'that's me!', Vuyanzi reinforces her idea that, even in subject choice, the student is a unique individual with their own personality. Such emancipatory discourse is essential in questioning cultural gender concerns.

But what about other views? Here are some thoughts from other students:

Boyi (provincial boys school): I first touched a computer in this school, but the other computer science students [already] had an idea. There are even other boys who have a computer at home and refuse to choose this subject. They already know what we are learning. I cannot afford to pay for private college, so I took this subject. I hope to be a doctor, and I know I will need a computer.

Boyi's readiness to choose computer studies can be understood as an intrinsic reward, measured in terms of the computer skills he will develop, not necessarily by educational 'outcomes'. Boyi, unlike Vuyanzi, does not see the subject of computer studies as a professional pursuit but as a relational influence for future careers. Considering his view about prior exposure, the following was noted about the content offered in computer studies:

Form One – Introduction to Computers

– Computer Systems

– Operating Systems

Form Two – Application Packages

– Word Processors, Databases, Spreadsheets

– Desktop Publishing, Internet, Email

– Data Security and Control

Form Three – Data Presentation in a Computer

– Data Processing

– Elementary Programming Principles

– Systems Development

Form Four – Introduction to Networking and Data Communication

- Applications Areas of ICT

- Impact of ICT in Society

- Career Opportunities in the Computer Field

- Project

Source: Ministry of Education, Science and Technology Secondary Education Syllabus, Vol. 4, September 2002.

The intention of the policy is clear. It revolves around exposure to the beginner, expressing the perceived imbalances in varying school environments, making the subject content 'user friendly' to girls and alienating them from the subject content. Differences may perhaps occur in the learning process. The following comments further illuminate the argument advanced in this paper:

> **Waridi (provincial girls school)**: When I joined this school, I was put into the computer science class. The teachers choose the top five girls in each Form One class …. I have done well so far and I hope to score… mmh… A in this year's exams. You have to be alert in this subject. Many of the girls find it is too demanding. The rest want to be secretaries in big NGOs, but I want to be a lawyer.

Waridi advances the view that the school policy, subject demands, gender stereotypes and personal ambition determine students' placement and success in the subject. On one level, she sees the key to doing well as being 'alert'. Such an idealised solution resonates with the deeply held psychological needs of girls that cement the gendered cornerstones of computer studies.

The influence of the family is also quite profound in some of cases:

Sawe (district boys' school): My parents want me to be an engineer, and my father says I cannot work without using a computer…

Kavuzi (district girls' school): I think what my mother says about this subject is true; it is important that I know this subject and get my marks right. I can only use a computer here in school, since at home we don't have one.

Ronika (provincial mixed school): The subject is good, and my teacher says so. But in class he always prefers boys answering the questions, and the boys never give us enough time on the computers, and we always have to fight for ourselves. But I will succeed!

Manyasi (district mixed school): *Wasichana* [girls] can only be good at typing and don't do very well in this subject. I know I will be a computer engineer, so I work hard.

Gender stereotypes can be clearly seen from the students' remarks. And as Ronika and Manyasi demonstrate, these stereotypes can be incorporated in the students' lives over time. Computer studies is seen rewarding by most of the students in career terms. The students' views structure any interpretation, especially in relation to the issue of agency, specifically parental influence and teachers' views. Cognitive accomplishment is important, but Ronika describes significant agents such as the teacher and male students that enhance gender differences and stereotyping picked up in the classroom. Many of these stereotypes are reinforced by students' previous computer knowledge. More than 80 percent of the boys in Forms One to Four who enrolled for computer studies had previously used computers compared to only 15 percent of the female students. These stereotypes are often cemented at home, where the boys are more likely to be encouraged to take an interest in computers (80 percent) than the girls (15 percent). Many girls reported that their brothers were allowed to leave the home and even given financial assistance for internet browsing, thus often isolating the girls from the world outside the class that uses computer technology. However, this view was not typical of the whole group, since some schools were in rural areas where electricity points were only found at schools and hospitals, thus limiting computer access outside school for both boys and girls.

Under these circumstances, differences in computer use may be detected either in or out of the classroom. Ronika and Kavuzi want to be successful and refuse to regard the subject as structurally masculine. Ronika seems aware that the boys see girls as unfit for computer science, but this only makes her resolve to work harder. In identifying teacher-gendered tendencies and masculinity in the classroom, she attempts to resist notions of gendering in the computer studies class. Nonetheless, such gendered notions exacerbate rather than alleviate gender disparities. In co-educational schools, more is involved than competition among learners, and the situation resonates with deeply held psychological needs concerning dependency, particularly given the patrifocal emphasis in most families. Manyasi's comment that 'girls can only be good at typing' reveals profound assumptions about gender, personal development, learning and teaching.

During the research the question, 'why have you opted for computer studies?', elicited a common general answer: 'because computer studies is necessary for success'. When probed further about what 'success' means, the learners said they needed careers based on computer science (70 percent male, 40 percent female) or careers where it would be an added advantage (50 percent male, 80 percent female). The latter explanation illustrates the same gendered assumptions about the link between subject choice, schools, families and individuals, forming a conceptual basis of understanding.

Running through the teacher's interviews were the gendered differences that arose out of cultural and situational contexts. Where the schools' computer studies project was supported by the community and PTAs were vibrant, these two agents had a positive social construction on computer studies. As was emphasised by one teacher, the expectations of the parents were that their children should not be left out of the computer world:

> **Teacher (provincial girls school):** Our parents are very supportive, but we must also consider the financial implications of building a computer room, the machines, etc. Some parents may want this subject, but remember it is an added financial

burden. With cost-sharing everywhere, we have to limit the subject to the very best students.

Such parental support for girls' educational programmes posits an ongoing tension between macro-structural pressures that increase the desirability of education for girls and micro-structural pressures that constrain girls' education in order to preserve a patrifocal cultural model of family. Computer studies acts as a discriminatory investment that does not benefit all; some communities cannot simply afford the added burden of computer studies. Not surprisingly, most students in 2002 came from the economically empowered provincial schools (51 percent), private schools (25 percent) and national schools (15 percent), while only 9 percent were from district schools.

Conclusion

Access to and experience with S&T plays a vital role in the career choice process. In the area of computer studies, gender assumptions and inequalities provide the structure for the ways we think about and value the teaching of computer studies. Whether our perception of the S&T gender gap is real or imagined, the research shows that girls are more interested in computers and other technology for their social function. The masculine view is more focused on the machine itself. In discussions and research findings, the influence of computer technology on culture and gender is often overlooked. A computer is not simply an instrument or tool but a component incorporated into a system (political, social, cultural, economic), and it is these systems that pose obstacles to gender and S&T. In a developing society such as that of Kenya, issues of gender and S&T development are pivotal in shaping the technological debate and informing feminist discourse.

Acknowledgement

The researcher would like to thank all the students, teachers and schools that participated in this project. The names given to students cited in this article are not their real names.

References

Beoku-Betts, J. A., 2003, 'Social Origins and Family Influences on Women Scientists in Third World Contexts', paper presented to the CODESRIA Gender Institute on Gender Science and Technology in Africa, Dakar, June 17-July 11.

Coughlin, P. and Ikiara, G., eds., 1988, *Industrialisation in Kenya: In Search of Strategy,* Nairobi: Heinemann.

Diakitie, S., 1991, 'Technological Rationality and Africa's Destiny', in K.K. Prah, ed., in *Culture, Gender, Science and Technology in Africa,* Windhoek: Harp Publications, pp. 117-127.

Enos, J. L., 1995, *In Pursuit of Science and Technology in Sub-Saharan Africa,* UNU/INTECH.

Erinosho, Y. S., 1994, *Girls and Science Education in Nigeria,* Abeokuta: Ango Intl. Publishing.

Galpin, V., 1992, *Gender and Computer Science Education.* (http://www.cs.wits.ac.za/vashti). September 24, 2003.

Green, F., and Adam, A., eds., 2001, *Virtual Gender: Technology, Consumption and Identity,* London: Routledge.

Grint, K. and Gill, R., 1995, *The Gender Technology Relation,* London: Taylor and Francis.

Harding, S., 1986, *The Science Question in Feminism,* Ithaca: Cornell University Press.

Harding, S., 1991, *Whose Science? Whose Knowledge,* Ithaca: Cornell University Press.

Harding, S. and McGregor, E., 1995, *The Gender Dimension of Science and Technology,* Paris: UNESCO.

Haraway, D., 1991, 'Situated Knowledges: The Science Question in Feminism and the Privilege of Partial Perspective,' in Haraway, D., ed., *in Simians, Cyborgs and Women: The Re-invention of Nature,* London: Free Association Books, pp 183-201.

Jegede, J. O., Agholor, R. and Okebukola, P., 1996, 'Gender Differences in Perceptions and Preferences for Social-Cultural Climate in Nigeria', *Research in Education: An Interdisciplinary International Research Journal,* Vol. 55.

Kakonge, E. W ., 2000, 'Gender Differences in Science Subjects in Secondary Schools in Kenya: An Investigation of Entries, Attainment and Teachers' Perspectives', unpublished PhD. thesis, University of Leeds.

Kitetu, C., 2003, 'Gender in Science Education', paper presented to CODESRIA Gender Institute on Gender, Science and Technology in Africa, Dakar, June 17-July 11, 2003.

KNEC, 2002, *Kenya National Examination Council / Kenya Certificate of Secondary Education, Entries and Performance in 1998 – 2002,* Nairobi: KNEC Data Dept.

Lukalo, F. K., 1999, 'Women and Educational Choices', paper presented at Annual Meeting, Association for World Education, RCEA Centre, Eldoret, Kenya.

Makau, B., 1999, *Measuring and Analyzing Gender Differences in Achievement in Primary and Secondary Education,* Nairobi: Academy Science Publishers.

Makhubu, P. L., 1993, 'The Potential Strength of African Women in Building Africa's Scientific and Technological Capacity', in *Science in Africa: Women Leading from Strength,* pp. 1-18. Washington, DC: AAAS.

Malcolm, S., 1993, 'Increasing the Participation of Black Women in Science and Technology', in Harding, S., ed., *The 'Racial' Economy of Science,* Bloomington: Indiana University Press, pp. 249-253.

Mariro, A., 1999, *Access of Girls and Women to Scientific, Technical and Vocational Education in Africa,* Dakar: UNESCO.

Nettleford, R., 2002, 'Africa, the African Diaspora and the New Millennium: Challenges for Development', in *Africa Today,* Vol. 8, nos. 11-12, pp 21-31.

Njenga, A. W., 1999, 'Promotion of the Equal Access of Girls to Scientific, Technical and Vocational Education', in A. Mariro, ed., *Access of Girls and Women to Scientific, Technical and Vocational Education in Africa,* Dakar: UNESCO, pp 163-182.

Obanya, P., 1999, *The Dilemma of Education in Africa,* Paris: UNESCO.

Olson, P., 1994, 'Feminism and Science Reconsidered: Insights from the Margins', in J. Peterson and D. Brown, eds., *The Economic Status of Women under Capitalism: Institutional Economics and Feminist Theory*. Aldershort: Edward Elgar.

Omoka, W. K., 1991, 'Applied Science and Technology: A Kenyan Case Consideration of their Interrelationship, in K. K. Prah, ed., *Culture, Gender, Science and Technology in Africa*, Windhoek: Harp Publications, pp. 14-31.

Sands, A., 1993, 'Never Meant to Survive: A Black Woman's Journey – an interview with Evelyn Hammonds', in Harding, S. ed., *The 'Racial' Economy of Science* Bloomington: Indiana University Press, pp 239-248.

Stepulevage, L., 2001, 'Becoming a Technologist', in E. Green and A. Adam, eds., *Virtual Gender* London: Routledge, pp. 63-83.

Truscot, K., 1994, *Gender and Education,* Oxford: Oxford University Press.

Wakhungu, J., 2003, 'Gender and ICT Development in Africa', paper presented to the Association of Third World Studies – ATWS, Kenyatta University, Nairobi, August 2003.

White, K., Shade, R. L. and Brayton, J., 2001, 'Lives and Livelihoods in the Technological Age', in E. Green and A. Adam, eds. *Virtual Gender*, London: Routledge, pp. 45-62.

Part III

Science and Technology:
The Case of One Woman, Many Women

11

Busy Career and Intimate Life: A Biography of Nahid Toubia, First Woman Surgeon in Sudan

Jackline K. Moriasi

Introduction

Not long ago, Dr Gerda Lerner said in her address as the new president of the Organisation of American Historians that, 'if the bringing of women—half the human race—into the center of historical inquiry poses a formidable challenge to historical scholarship, it also offers sustaining energy and a source of strength' (Lerner 1982:69).

Women have contributed in every possible way to the technical advancement of humanity. They have carried the same burdens of scholarship as men, and they have accomplished just as much. They have been as resourceful and passionate about their work, and certainly as creative. Such women have left a remarkable legacy. Their stories are a clear light to the future.

Yet, women remain grossly under-represented in science and technology professions worldwide, especially in developing countries. This fact has generated considerable concern, mainly among feminists but also in the society at large. Feminists have stressed the need for equality of opportunity for both sexes and the desirability of reducing the existing male domination of science and technology. Society recognises that it is essential to harness the vast human resources of women in order to promote development (Harding 1998). Thus, the under-representation of women in science and has serious practical implications. Failing to encourage women to develop and use their talents in science and technology amounts to a serious drawback, considering women's numerical strength and the abundance of their scientific potential. The potential of women to utilise their scientific capacity as much as men is evident in the life of Nahid Toubia, the first woman surgeon in Sudan.

It has often been argued that oral history is a basic tool in our efforts to incorporate the previously overlooked lives, activities and feelings of women into our understanding of the past and the present. When women speak for themselves, they reveal hidden realities; new experiences and perspectives emerge that challenge the 'truths' of official accounts and cast doubt on established theories. Interviews with women can explore private realms such as reproduction, child rearing and sexuality to tell us what women actually did, can do or should have done. Interviews can also tell us how women felt about what they did and can reveal the personal meanings and values of particular activities. They can, but they usually do not (Gornick 1983). Women have much more to say than we have realised. As oral historians, we therefore need to develop techniques that will encourage women to say the unsaid. It is from this perspective that I trace the life and times of Nahid Toubia.

Nahid Toubia: From Humble Environment to International Fame

Nahid Toubia was born in Khartoum in 1951 and she attended medical school in Egypt to become the first woman surgeon from Sudan. In 1981, she became a fellow of the Royal College of Surgeons of England and the first woman surgeon practising in Sudan, serving as the head of the Pediatric Surgery Department at Khartoum Teaching Hospital for many years. Recently, she worked for four years as an Associate for Women's Reproductive Health at the Population Council in New York City. She is currently an Assistant professor at Columbia University School of Public Health and the founder and director of Research, Action and Information Network for Bodily Integrity of Women (RAINBO) in New York. She is a member of several scientific and technical advisory committees of the World Health Organisation, UNICEF and UNDP and vice-chair of the advisory committee of the Women's Rights Watch Project of Human Rights Watch, where she serves on the Board of Directors. She is also member of the Board of the Association for Voluntary Surgical Contraception (AVSC) International.

Barriers to women deriving from the structure of the academic system are reinforced by 'cumulative disadvantage' factors that exclude other women from science but also carry over and affect the academic careers of women. These include the differential socialisation of men and women, impaired self-confidence and negative expectations regarding the impact of children on women's academic careers. The roots of this problem lie in the different gender experiences of boys and girls. As young girls and women, females are socialised to seek help and be help givers rather than to be self-reliant or to function autonomously or competitively. Girls are encouraged to be good students in so far being given a task, completing it well and then receiving a reward from an authority figure.

However, in graduate school, behaviour is expected to be independent, strategic and void of interpersonal support. These expectations are antithetical to traditional female socialisation. In addition, the needs of women, based on socialisation which encourages supportive interaction with teachers, is frowned upon by many male and some female faculty as indicative of inability. As one female graduate student put it, 'the men have the attitude of "Why should people need their hands held?" ...' Lack of a supportive environment exacerbates an often already low level of self-confidence.

Assembling Women's Perspectives

Recent feminist scholarship has been sharply critical of the systematic bias in most academic disciplines, which have been dominated by the particular and limited interests, perspectives and experiences of white males. Feminist scholars have insisted that the exploration of women's distinctive experiences is an essential step in restoring the multitude of both female and male realities and interests to social theory and research.

Assembling women's perspectives is therefore important for feminist scholars because women's experiences and realities have been systematically different from men's in crucial ways and need to be studied in order to fill large gaps in human knowledge. This reconstitution of knowledge is essential because of a basic discontinuity. Women's perspectives have been neglected not simply as a result of oversight; they have been systematically suppressed, trivialized, ignored or reduced to the status of gossip and folk wisdom by the dominant research traditions institutionalised in academic settings and in scientific disciplines. Critical analysis of this knowledge often showed that masculine biases lurked beneath the claims of social science and history to objectivity, universal relevance and truth.

The need to study women's realities and perspectives raises methodological as well as substantive issues. Dominant ideologies have distorted and made invisible women's real activities, to women as well as to men. For example, until recently, it was common for women to dismiss housework as 'not real work'. Yet, unlike most men, women also experience housework as actual labour, as a practical activity that can fill their daily existence. In effect, women's perspectives combined two separate consciousnesses: one emerging out of their practical activities in the everyday world and one inherited from the dominant traditions of thought. Reconstructing knowledge to take account of women, therefore, involves seeking out the submerged consciousness of the practical knowledge of everyday life and linking it to the dominant reality.

The perspectives of two feminist scholars, Marcia Westkott and Dorothy Smith, have especially influenced our thinking about oral history. Westkott provides us with a basic approach to individual consciousness. She describes how traditional social science assumes a fit between an individual's thought and action 'based on the condition of freedom to implement consciousness through direct activity'. But within a patriarchal society, only males of a certain race or class have anything approaching this freedom. Social and political constrains have limited women's freedom. Thus, in order to adapt to society while retaining their psychological integrity, women must simultaneously conform to and oppose the conditions that limit their freedom.

In order to understand women in a society that limits their choices, we must begin with the assumption that what they think may not always be reflected in what they do and how they act. Studying women's behaviour alone gives an incomplete picture of their lives, and the missing aspect may be the most interesting and informative. Therefore, we must also study consciousness, women's sphere of greatest freedom, and go behind the veil of outwardly conforming activity to understand what pariticular behaviour means to women and, reciprocally, how women's behaviour affects their consciousness and activity.

Sudan: A Geo-Political Background

Sudan, the home of Nahid Toubia and the place where she was nurtured and brought up, is divided between the north and the south over issues of religion, political structures, culture and race, differences that have greatly intensified since the military coup of 1989. In that year, the series of ineffective coalition governments led by Prime Minister Sadiq al-Mahdi were finally brought to an abrupt end when army elements led by General Omar Hassan Ahmed al-Bashir took power. The subsequent desire to enhance Muslim rule and practices in the south of Sudan led to outright rebellion there.

Women constitute approximately 15.6 million out of a total Sudanese population of 31.6 million. Women play a key role in the economic field, with females constituting 26.5 percent of the labour force. This is up from only about seven percent of the work force in the 1960s. Article 21 of Sudan's 1998 constitution clearly states that all Sudanese are equal before the law without discrimination as to sex or race. All labour legislation is based on complete equality between men and women. The 1998 Constitution reiterated and reinforced earlier equal employment opportunities clauses in the 1973 constitution. These provisions were reinforced in the 1997 Public Service Act, which provided for equal wages for equal employment, open competition based on competence, qualifications and experience, equal pension rights and equality regarding leave and holidays.

In November 2000, the president decreed that women would receive two years paid maternity leave. While most women work within the agricultural sector, a large percentage also work as professionals, serving as ambassadors, university professors, doctors, lawyers, engineers, senior army officers, journalists and teachers. There are, for example, women major-generals in the police. In 1996, the United Nations Economic Commission for Africa published a book entitled 'Africa's Roll of Distinguished Daughters'. Of the fifty African women listed, ten were Sudanese; and these included academics, lawyers, journalists and psychologists. Politically too, women are well represented.

Sudanese women became involved in national politics from the mid-1940s onwards and secured the right to vote in 1953. In Sudan, women have an unfettered right to elect and be elected in presidential, federal, state and local elections. To offset innate conservatism and to ensure female participation in political life, there is a quota system guaranteeing female seats and participation in federal and state legislatures. A quarter of all federal parliamentary seats are reserved for women. Women are also ensured a minimum of ten percent of seats in all other elected local bodies. Women have chaired select committees within the federal National Assembly, and there have been women ministers in Sudanese governments since the early 1970s. There are several women ministers in the present government, holding portfolios such as health, social welfare, public service and manpower and cabinet affairs. In 2000, the Sudanese President appointed a cabinet-level Advisor on Women's Affairs. There is also an advisor within the Southern States Coordinating Council. There have been, and are presently, women ministers within several of the state governments. There is women's policy unit within the Ministry of Social Planning, drawing up national policies and plan's for women's development.

In the field of education, which is the linchpin of development, the state has adopted universal and compulsory primary education and achieved remarkable success in the field of general education, where both girls and boys are enjoined to be in school with government support. The percentage of female intake in the period 1990-1998 increased by 22 percent, whereas the boy's intake increased 8 percent. Sudan witnessed a tremendous increase in girls' enrolment in secondary schools to 75 percent during the period 1993-1998. The state has also given considerable recognition to higher education and special attention to science and technology fields. The percentage of female students in universities increased to 62 percent by 1999, compared with 47.2 percent in the year 1995.

Nahid Toubia's Education

Nahid went to nursery, primary and intermediate school at a local church school that, at the time, was apparently called the American Mission, but was later named the Evangelical School for Girls. She then decided to change from the private school system, which gave very light education for girls that did not qualify them for university, to the state system, which gave girls a serious education. She sat the state exams and was admitted in Khartoum Secondary School for Girls, at the time one of the two top girls' schools in the country (Personal Interview, 2003).

Nahid joined the University of Khartoum in 1968 for her first year in premedical science. Between 1969 and 1974, she spent five years in medical school at the University of Cairo's Qasr el Aini medical school, the oldest and most prestigious medical school in Egypt. She graduated as a medical doctor in 1974 at the age of 23. After practical training at Khartoum Teaching Hospital, she went to England for a degree in surgery. She received her Fellowship of the Royal College of Surgeon of England (FRCS) in 1981. Nahid was 29 years when she became the first woman surgeon in Sudan. In 1985, she started a private clinic in Khartoum where she practised general and pediatrics surgery and was very successful in the private as well as the public system. After 15 years of practising surgery, during which she went into pediatric surgery and became the head of the Department of Pediatric Surgery at Khartoum Teaching Hospital, Nahid left the clinical field and returned to the University of London for a degree in Health Planning and Financing, which she obtained in 1989.

Nahid explains that she was prompted to go into the sciences as a result of hard work and being what she calls 'inquisitive'. She says:

> I always had an inquisitive mind, but I never liked the abstractness of Maths. In high school, I was excellent in biology and English. There were two tracks; one had to choose either sciences or arts and humanities. Science was for the smart pupils, and arts and humanities were for the less smart and for girls who just wanted to have an education but not a career. I never contemplated going into arts. My self-image was of a smart and a professional person (Personal Interview, 2003).

From my interview with her, it became clear that Nahid's mother always had a very high opinion of doctors. She saw the profession as the ultimate in human service and had a family friend who was a model of a caring doctor and was her idol. Nahid was her mother's daughter and wanted to be her idol, so she did not have any family problem in choosing what she wanted. Her family was very liberal for their time. Her father believed girls should be allowed the same freedoms as boys. In fact, sometimes he was harsher to her eldest brother, as he thought boys were more likely to be led astray. Her mother believed girls and women could do anything they set their mind to. She was also the handy person in the house and fixed everything from dolls to leaking taps. Nahid was named after a well-known Egyptian suffragette from the women's movement of that time, Nahid Sirri (Personal Interview, 2003).

The only problem Nahid encountered in choosing science is that the intermediate school she attended was a church-affiliated school whose standards in maths and sciences were not very high. She therefore realised that, if she continued in that school, she would never have a chance of going to the university. She was only 12 years old, but she made her decision to change from that school. The decision meant that she had to study alone and take extra maths lessons to sit for the state exams. Her family was supportive of whatever she wanted. They did not interfere but trusted her judgment, as she was a very serious child. She sat for the exams, passed and went on to the best government school for girls, where she was able to get the education that got her admitted in the university, in a science faculty (Personal Interview, 2003).

From Classroom Science to Pragmatic Science

Nahid's career has been very mucuh concerned with reproductive health, especially concerning female genital mutilation and women's human reproductive and sexual rights. She has been on the editorial committees of *Health and Human Rights*, a journal of the Harvard School of Public Health, *Reproductive Health Matters*, an independent journal published in London, a member of DAWN (Development Alternatives with Women for a new Era), a member of the advisory committees of the Global Fund for Women and the Tropical Disease Programme (Edna McConnell Clark Foundation), a vice-chair of the advisory committee of the Women's Rights Project of Human Rights Watch as well as a member of the board of directors of

Human Rights Watch and a member of the board of trustees of Association for Voluntary Surgical Contraception (AVCS). Some of her many publications include *Caring for Women With Circumcision: A Technical Manual for Health Care Providers* (RAINBO 1999a), *Learning About Social Change: A Research and Evaluation Guidebook, Using Female Circumcision as a Case Study* (RAINBO 1999b), *Sexual Coercion and the Reproductive Health of Women* (Population Council 1995), *World's Women: Trends and Statistics* (United Nations 1995) and *Arab Women: A Profile of Diversity and Change* (co-authored with Amira Bahyeldin, Nadia Hijab, and Heba Abdel Latif) (Population Council 1994).

Nahib has translated her scientific prowess into practical work by participating in the formation of RAINBO, an international NGO working on issues at the intersection of health and human rights of women. She also established the AMANITARE programme, which evolved from the need for a coordinated pan-African effort to consolidate the skills, knowledge and institutional resources of groups and individuals active in the field of sexual and reproductive health, gender equality, and women's rights. It is an effort to facilitate the translation of the principles embodied in these agreements into the daily realities of African women and girls. It aims to create a better political, economic and social environment to enable African women and girls to enjoy their lives without fear of control or coercion because of their sexuality or reproductive potential.

Nahid has also translated her applied scientific knowledge to the development of African youth. This is well illustrated in her fight against clitodectormy or Female Genital Mutilation (FGM). She has published and written many articles on the subject. In this section, I will try to highlight her views on female circumcision. According to Nahid, FGM is practised in twenty-eight African countries on anywhere from 5 to 95 percent of those countries' young women. Asked why she had chosen to approach FGM primarily as a human rights violation, rather than as a threat to women's health, she replied:

> At RAINBO, the health and human rights aspects go hand in hand. For instance, by working on it solely as a health issue, it loses a very important point. If the girl experiences no health complications and the procedure went very smoothly, we still have to face the fact that a major human rights violation has occurred. A person, a girl, has had a very sensitive part of her genitalia permanently removed and she has had no real power to stop it.

Viewing FGM as a rights issue links it to its core problem, that is, the status of women. Nahid argues that men do not go through anything similar, such as the amputation of the penis, because they happen to be in positions of power. She therefore argues that seeing FGM in a rights framework is getting to the core of the *solution*, as well as the problem. As to why parents do not resist the practice, she observes that parents have to live as social beings and conform to the social norms of their environment. We all do a lot to be accepted in our societies, and the more dependent we are on that society for our survival, the more willing we are to conform to that society's rules.

In Nahid's thinking, even women themselves have participated in perpetuating patriarchy. For hundreds of years, most women in most parts of the world have accepted their second-class status. International actors play a major role at this point because they have the funds and the political power to influence government policy and to help NGOs technically and financially. But they are also limited. Social change can only happen from within, and no outsider, not even other NGOs from a neighboring community, can single-handedly affect change in a social environment where they are the outsiders.

Talking about international law, but, when you have individual girls who have been circumcised and they want some redress, or their parents want redress, you have to give them a tool by which to get it:

> I cannot emphasis enough that the only way FGM will be abandoned is through fundamental social change. Laws and policies are a factor in the dynamic of social change. You inform people of the harmful effects of FGM. You appeal to their concern for their daughters. And in the push and pull of change versus status quo, the law becomes a pushing force. It gives support to those who want to change, especially those less powerful in society (Personal Interview, 2003).

Such laws, she notes, have only been around for a very short period of time, nor have we invested enough in research. Thus, anything we say now is impressionistic and anecdotal. The effect of a law varies tremendously depending on the social and political context. If a law is passed in a country where there has been a lot of activism to prepare the ground, it may have a very different effect than in a place where the issue has barely been mentioned. Regarding the effect of such laws, Nahid says, 'it is difficult to assess but our hope is that these laws will empower women and girls and not increase their vulnerability'. Thus, she recommends that the laws should be geared toward practitioners, not toward the girl child or the parents. Commenting on the short-term and long-term complications and implications of FGM, she argues:

> It all depends, of course. People like to exaggerate, this sort of 'Oooh! Ahhh!' of the bleeding, the infection, this, that. Of course, all that happens. It depends on the degree of cutting. But there are also the psychological problems of children who might be traumatised by it, although not all children are traumatised by it. Sometimes, they perceive it as a positive experience, even though they somehow suppress their fear of it. It's quite a complex psychological process. Also, its effects on sexuality tend to vary quite a bit, depending on the severity and also the social constructs of sexuality in any particular place.

Nahid Toubia thus sees FGM within the framework of its cultural significance. Although she admits that, 'if you ask people immediately why they do it, most people would just tell you, 'Well, it's just a tradition, we have not thought about why we do it', her view is that the practice is responsible for the suppression of women's sexuality or interfering with it in some way. All in all, Nahid's scholarly work and engagement with societal problems portray her as a strong role model for African girls.

References

Harding, S., 1998, *Is Science Multicultural?* Bloomington: Indiana University Press.

Gornick, V., 1983, *Women in Science: Portraits from a World in Transition,* New York: Simon and Schuster.

Lerner, G., 1982, 'President's Address to the Organisation of American Historians', in *Journal of American History*, Vol. 69, no. 1, pp. 7-20.

RAINBO, 1999a, *Caring for Women With Circumcision: A Technical Manual for Health Care Providers.*

RAINBO, 1999b, *Research and Evaluation Guidebook, Using Female Circumcision as a Case Study.*

12

Assessing the Impact of Coffee Production on Abagusii Women in Western Kenya: A Historical Analysis (1900–1963)

Samson Omwoyo

Introduction

African agriculture has experienced drastic changes in its organisation and form, thus affecting women, the key players, in diverse ways. Both internal and external factors have contributed to this transformation, including the introduction of technologies and innovations, such as new crops. Nevertheless, agricultural performance on the continent has worsened, and, although this generalisation tends to ignore changes in specific and small units of analysis, Africa is largely portrayed as a continent plagued with endemic food shortages and famines. Women, especially in rural districts, who spend much of their time and labour on agriculture, have suffered greatly from such poor performance.

This chapter takes a small unit of study, the Abagusii community in western Kenya, and sets out to analyse historically the changes in agriculture over a fairly long period of time, from 1900 up to 1963, when Kenya attained independence. It is evident that Kenya's poor agricultural performance has been a culmination of processes and changes brought about by colonial capitalism. The internationalisation of division of labour relegated Africa to the role of supplier of cheap agricultural raw materials to the Western capitalist world. The integration of Africa's economy in general, and Kenya's in particular, into the world capitalist system elicited a process of transformation that gradually modified, marginalised and subordinated the region's agriculture, with severe implications for women, who are the key players in the sector. Thus, colonialism effected structural changes in African agriculture that have weakened both African food production and the role women play in it. This chapter takes one agricultural innovation, coffee farming, and analyses the effects it had on women's role in agricultural production among the Abagusii of western Kenya during the colonial period.

Women have always been an integral and crucial component in agricultural production in Kenya. However, up to the 1980s, most studies on Kenyan agriculture either totally ignored or dealt scantily with women's role. African peasant farmers were not differentiated along gender lines, thus subsuming and obscuring female agricultural producers. Yet, for example, women in Kenya comprise about 52 percent of most Kenyan communities, the Abagusii included, and most of them live in rural areas where agriculture is the predominant occupation. Thus, studies concerning agriculture and food production that do not relate to women are incomplete. Moreover, among the Abagusii, it is women who have been primarily responsible for food production, household management and the nurture of children (Stichter 1982).

Since the 1980s, however, studies have begun to pay serious attention to how socio-cultural and economic changes in agriculture affect women as well as men in Africa (Johnson and Kelb 1985; Meillassoux 1981; Boserup 1970). The role of women has generated much interest and research, and an enormous literature on gender has emerged focusing on the role of women in society generally. Nevertheless, numerous gaps remain to be filled, especially in historical studies. Moreover, due to the patriarchal nature of Kenyan society, most women neither own the means of production nor control the proceeds from their labour. The exploitative nature of gender relations in agriculture has survived from the pre-colonial era. Historical analysis can therefore elucidate the precarious position women have had to endure over time, while also showing that they were agents of their own destinies as they struggled to cope with the new changes (Olson 1994).

This chapter blends two important themes with a view to critically analysing the changing role of women in agricultural production in the face of agricultural innovations, in particular, the introduction of coffee among the Abagusii. My premise is that the role of women in agricultural production was disadvantaged in the pre-colonial period, given male dominance of the factors and relations of production. Men used the prevailing patriarchal social and economic relations to appropriate women's surplus, but this exploitative relationship was amplified by the technological innovations accompanying the introduction of coffee in the area. The lucrative cash crop was often the preserve of male farmers, while female farmers were relegated to subsistence farming. Ongoing research into high-yielding varieties of coffee, along with the use of pesticides and fertilisers, benefited the male farmers most, thus entrenching their dominant role in 'modern' agriculture, often using female labour for the benefit of men. Where labour migrancy occurred, usually to the neighbouring tea estates in Kericho area, female farmers were further burdened, as they had to take up roles formerly performed by men.

Such changes to women's participation in agricultural production, occasioned by the introduction of coffee in Gusiiland, are the concern of this chapter. The response of women in coping with their continued marginalisation, and their methods of survival will be analysed taking into account the fact that Gusii women do not constitute a homogenous class. Nor were they equally and uniformly affected by the changes. They had different amounts of land, different levels of education, different family sizes and different numbers and ages of children. Local climatic and soil conditions, and many other specific conditions, differed from area to area or farm to farm. Nevertheless, the key point is that women's efforts to resist and change their marginal position in coffee-farming portray them as agents of their own destiny rather than victims of male patriarchy and dominance.

This chapter, therefore, endeavours to achieve several objectives. First, it examines the impact of patriarchal relations on the role of women and men in agricultural production in pre-colonial Gusiiland. Secondly, it analyses the impact of colonialism and the introduction of coffee on the role of women and men in agricultural production in Gusiiland. Thirdly, it investigates the responses of women to their continued marginalisation; and, finally, it tries to identify the factors that influenced these responses and impacted on the role of women in agricultural production.

Capitalism in Patriarchal Relations

The approach of analysing capitalism in the context of patriarchy has been used as an important way of examining women's participation in agricultural production in Kenya. This approach encompasses patriarchal relations, the articulation of modes of production and feminist standpoint theories. Patriarchy basically denotes the role of the father as the 'head' of the family (Lerner 1986), but it also describes the political, economic and social control of women by men. Men are thus perceived as decision-makers, especially at the public level, while women are relegated to the periphery and viewed as inferior and subordinate to men. At the household level, men are branded the 'breadwinners', while women are seen as mere recipients together with other members of the family. The male head of the household during the colonial era paid the hut and poll taxes for every member of the homestead. This patriarchal ideology solidified during the colonial era, as the colonialists relied on the Victorian ideology of the woman as good/responsible housewife/lady while providing jobs to male loyalists. Over the years, patriarchy has valorised the dominance of men over women and empowered men to exercise male authority absolutely, to the extent that they come to be perceived as 'natural' leaders, both in the household and in the wider society. In the realm of agriculture, women are often the sole producers, but the proceeds go to the men in their role as heads of the families. Again, some new technologies that came in the agricultural fields are mostly taught to women, as they do the bulk of the agricultural work. Despite this, the land in which these technologies are used belongs to men, as they are the inheritors of ancestral land. The profits that accrue from the new technologies thus go to the men, to the detriment of the women.

On the other hand, a mode of production is seen here as a system of production or social form of economic organisation. It mainly involves itself with the means of production and the attendant social relations of production. The main argument in this theory is that, when the capitalist mode of production is introduced, it does not automatically and immediately replace the pre-capitalist modes of production but, rather, reinforces them. With time, the capitalist mode of production gradually asserts itself over the pre-capitalist mode of production, and the two modes of production are then locked in a complex and sometimes contradictory struggle. Gradually, the capitalist mode of production modifies, marginalises, or subordinates the pre-capitalist mode of production, but by utilising it rather than casting it aside. The pre-capitalist mode of production is not completely eliminated but keeps on reproducing itself diversely in relation to the capitalist mode of production. Goodman and Redcliff (1981:60) thus observe that pre-capitalist modes of production may have continued to exist, though subordinated to the capitalist system, through a process of 'preservation and destruction' or 'dissolution and conservation', by which they were articulated in their diverse relations with capitalist system, particularly through unequal exchange relations.

Articulation is therefore a double-edged concept where certain sectors of the pre-capitalist economy were integrated into the capitalist economy and other sectors were not integrated for some time, with a view to achieve certain economic goals. This explains why men readily embraced the new agricultural technologies, including the introduction of coffee, and thus belonged to the 'modern' sector while relegating women to the backwaters of the economic realm in the name of the pre-capitalist sector, also called the subsistence sector. Just as the capitalist mode of production preserved the pre-capitalist sector in order to utilise and exploit it, so did men in relation to women. Through patriarchal relations, men kept women in the pre-capitalist sector so as to use and exploit them. Men grew cash crops, while women grew subsistence crops, but even then men utilised women's labour in all their endeavours. Thus, the theory of articulation of modes of production can aptly be applied to explain why women's participation in agricultural production has been hampered and thwarted to meet men's capitalist objectives.

However, women were not passive recipients of the changes affecting them. Nor did they all respond in the same way or get affected uniformly. To appreciate such variations, I use feminist standpoint theory. This theory seeks to interrogate different situations under different conditions so as to arrive at results that are differently conceptualised. It assumes that phenomena are differently located materially, socially, politically, economically and culturally according to the interplay of various factors. Thus, by taking into account the multiplicity of factors affecting a phenomenon, one arrives at conclusions that are only tenable to the particular condition. In other words, one expects different results from common stimuli depending on the specific conditions on the ground. Universalisation and homogenisation are thus eliminated (Harding 1998; Olson 1994; Goonatilake 1984). This theory proves useful in discerning why and how women responded differently to the agricultural changes.

The research for this paper was based on documentary sources of information. Primary information was obtained from archival records, which yielded useful in-

formation on the colonial era. Secondary information was sourced from various
libraries in Kenya and from the internet. The data obtained was corroborated to
ensure validity, then analysed qualitatively and descriptively. The findings are presented below.

Pre-Colonial Patriarchal Relations

The pre-colonial Gusii were mixed farmers, herding animals such as cattle, sheep
and goats, as well as cultivating crops such as *wimbi* (finger millet) maize, *mtama*
(sorghum), pumpkins, sweet potatoes and cassava. Women played a crucial role in
the agricultural process. For example, land that was the fundamental resources for
crop cultivation and animal production was designated into different uses with women
in mind. The arable land was basically divided into three main parts. The first was
land on which a family homestead was located and on which the wife, or wives) the
Abagusii being largely polygamous) carried out farming. The second was the land
the patriarch cultivated for his private use or as security in case of food shortage.
This, too, was divided out for each wife, although the produce was considered the
patriarch's. The third portion, consisting of all the remaining land, was communal
and belonged to the clan. Women could gather fruits, firewood, vegetables and
medicinal plants here as they wished. It has been found that the survival of humankind has been due much more to 'woman-the-gatherer' than to 'man-the-hunter'
activities, thus making women's productivity the precondition of all human productivity (Mies 1998:58).

Women were vitally involved in agriculture, but, either as individual households
or group parties, they participated in what emerged as a vertical, unequal and hierar-
chical sexual division of labour (Mies 1998:48). In the preparation of land for
cultivation, for example, women did the digging in two stages: first, just tilling or
breaking the ground and, second, pulverisation, involving the collection of all veg-
etative matter, which was then heaped in moulds called *amatuta*. The sowing of
wimbi and planting of many food crops was the preserve of women. Women and
their dependants largely did the weeding and harvesting of wimbi and other crops.
In all these processes, women exhibited a mastery of agricultural knowledge, including identifying fertile areas to be cultivated, selecting good seeds for sowing and
inter-cropping to minimise labour and maximise output. Their expertise even extended to designing ideal storage devices that minimised wastage and loss through
rotting or exposure to moisture, pest infestation or attack from animals.

Women were also the household custodians of food. All the food requirements
within their households were their concern. However, although they were in charge
of the food they produced and were free to sell or exchange any surplus, these were
limited areas of semi-autonomy; on the whole, they were still subjected to patriarchal dominance and exploitative relations by men (Mies 1998:61). The enterprising
women among the Gusii, for they were not a homogeneous and undifferentiated lot
(Olson 1994), acquired livestock of their own besides what was apportioned to
them by the head of the homestead. One cardinal principle among the pre-capitalist

Gusii was that there was no direct payment for labour contributed in many domestic or agricultural activities. Labour was mainly compensated for in kind, for it was held that to pay meant that one had not been truly assisted. It is in this vein that women's participation in agricultural activities should be seen. They variously participated, from the household level to the communal level, through such forms of cooperative labour parties as the *egesangio, ekebasono* and *risaga*. These were inter-household forms of group labour by women and girls who helped one another in tasks, such as weeding, on a rotational basis. The parties were seasonally formed and dissolved after the need for them was over. Due to the ongoing subsistence production of women, the men were free to go from time to time on hunting expeditions, which were sporting and political activities rather than an economic one. Mies (1998:58) calls hunting 'an economy of risk' and argues that 'the various forms of productivity which men developed in the course of history could not have emerged if they could not have used and subordinated the various historic forms of female productivity'. It is evident, therefore, that women played a crucial role in agricultural production among the pre-colonial Gusii.

Imposition of Colonial Rule on Gusiiland

By the Anglo-German Treaty of 1890, Gusiiland fell under the British 'sphere of influence'. By 1903, an administrative post had been established at Karungu on the shores of Lake Victoria, with an Acting District Commissioner in charge. The otherwise 'independent' Gusii eventually had to be subdued in a 1905 punitive expedition, and a permanent administrative post was then established in the present-day Kisii town. Soon the Gusii started paying taxes and offering their labour to the colonial order. Any lingering resistance to the colonial order was quashed with the defeat of the 1907 and 1914 uprisings.

Colonialism is a system of administration, a process of exploitation and a production system geared towards the creation of capitalist relations and the economic and socio-cultural aggrandisement of the coloniser. It involves covert and overt psychological, legal and military mechanisms (Emeagwali 2003). Thus, the penetration of colonial capitalism threw the Gusii pre-colonial economy into disequilibrium, and the Gusii found themselves subject to an economy over which they had little control (Omwoyo 1992:65). Gusii farmers, many of them women, started producing surplus for sale in order to pay taxes imposed on the male patriarchs. Obviously, the definition of what is 'necessary' and what is 'surplus' is not a purely economic question, for, as Mies points out, colonial exploitation is not only the one-sided appropriation of the surplus produced over and above the necessary requirements of other communities. This concept of exploitation, therefore, always implies a relationship created and maintained, in the last resort, by coercion or violence. Gradually then, the role of women as agricultural commodity producers was intensified through coercive and compelling circumstances, as the Gusii were forced by the colonial system to start growing crops for sale over and above the level of precolonial production. They were gradually introduced into the money economy and found themselves producing increasingly for sale (Ochieng' 1974:86). Consequently, the pre-colonial practice of selling the surplus was superseded by conscious production of surplus for sale, and a Gusii peasantry began to emerge.

Alhough, at first, production was mainly of indigenous crops such as *wimbi*, the colonial government was soon experimenting with other commodities. Gradually *wimbi* was replaced with improved hickory maize as the major crop, and maize soon became the major cash crop among the Gusii, as it had a ready market among white planters and settlers who needed to feed their workers. Maize was also easier and cheaper to grow; despite its low nutritive value in comparison to indigenous crops like sorghum and millet, it provided a greater quantity of food and the necessary energy that was needed for large numbers of workers. Maize made the workers feel more satisfied and well fed, so that they did their work with cheerfulness and vivacity. Maize thus grew to attain the status of monoculture even at the height of the introduction of coffee among the Gusii. Women, of course, continued to give much of their labour in maize production. The working parties were now focused on maize production, but women's involvement was set to be further intensified with the introduction of coffee at a time when many men were forced to work outside the district in the 1920s. Under capitalist patriarchal production relations, women were relegated to the 'subsistence' economy to free men for the colonial and capitalist sector.

By this time, tea plantations were being established in neighboring Kericho District. Tea is one of the most labour-intensive of all crops, requiring very large numbers of workers all year round. Due to its proximity, Gusiiland became the reliable labour reservoir for the Kericho plantations, and Gusii households as units of production, consumption and reproduction were radically altered in the process. Under the migrant labour system, men were drawn or forced off the land, leaving behind their women to maintain production. The costs of reproducing, maintaining and sustaining the cheap labour force were, therefore, borne by this 'pre-capitalist' sector run by Gusii women (Stichter 1982: Zeleza 1987). As the tasks and roles performed by men were changed, the workers' families remained at home, shouldering most of the burden of land cultivation while also suffering the imposition of forced labour in communal undertakings for the colonial authorities.

The costs of household production—including retirement, education, health and the rearing of the next generation of workers—were borne by the economy of the African 'reserves', which supported the workers' wives, children and themselves in sickness, old age or on leave. In this way, the pre-capitalist economy, a major preserve for women, became an appendage to the economy of estate agriculture, subsidising its low wages. In other words, women were invisibly exploited not only to keep the wages of the men in the estates sector low, but also to maintain their households with little need for productive input from the men. The men lived in an economic system based on women's productive agricultural work; they were the husbands of female agriculturists (Mies 1998:64). Household relations of production were also modified in varying ways, either in the direction of capitalist exchange or through the intensified exploitation of traditional obligations in the service of the labour market. Women became burdened with more work in the field. They also had to take up roles formerly done by their absent or migrant husbands. Thus wives' obligations to their husbands were intensified, as they were pressed to take over more work on family land-holdings (Stichter 1982:28). Against this already tilted and burdensome position of women, it is interesting to observe how the adoption of coffee production worsened the position of Gusii women still further.

Women's Increased Participation in the Technology of Coffee Production

For a long time, Kenyan Africans were barred from producing coffee by the colonial administration. It was claimed that Africans could not master the technical knowledge required to produce such a lucrative crop and would increase the risk of plant disease and inadequate quality control (Garst 1972:125). In districts near settler farms, it was said that the African plants would 'infect' settler coffee. The actual reason, of course, was the fear that African coffee-farmers would become self-sufficient and unwilling to offer their cheap labour on settler farms (Omwoyo 2002).

However, in 1934, Africans living in areas away from settler farms were allowed to grow coffee for experimental purposes, notable in Kisii, Embu and Meru areas. In Gusiiland, 64 beds of coffee seedlings were availed for planting. In the initial period, the crop proved unpopular with the Gusii. First, the period of care before yields were obtained was too long. Secondly, the colonial administration permitted that coffee to be grown on a cooperative basis only, in the hope of controlling quality and diseases. However, without personal commitment, farmers tended their plots irregularly. At times, the distance to the plots was a hindering factor. In fact, such farms had to be maintained under threat of prosecution, and for this reason, even when it was realised that individual plots near the farmers' homes could achieve more success, most Gusii were convinced that the government would confiscate their plots if they planted coffee. Consequently, as Barnes (1976) shows, a positive response to the introduction of coffee was forthcoming from only a small number of cultivators.

The chiefs, along with a significant number of the early-educated members of Gusii society, were among the first growers. They were motivated by a combination of reasons, including the expectation of earning greater cash income. By 1936, a total of only 50 acres owned by 25 growers had been planted in the Gusii highlands, with Chief Musa Nyandusi of Nyaribari having more than eight acres alone (KNA/DC/KSI/1/4/1937). However, by 1937, the attitude of peasants had positively changed in favor of coffee growing. Writing in 1937, the agricultural officer remarked: '…it is no longer a question of persuading people to plant but of selecting the most suitable applicants and allowing then to plant small areas only' ((KNA/DC/KSI/1/4/1937). In that year, the total acreage under the crop increased to 78, and, in December, the first parchment coffee, all from Chief Musa's farm, was dispatched to Nairobi for grading and sale. The reports on this parcel were encouraging—it was classed as borderline for the London Market—and, after this, there was even more demand for the local peasants to be allowed to grow the crop. By the end of 1938, 160 peasants were growing coffee on 90 acres (KNA/DC/1/4/1938). However, a maximum individual acreage had been imposed, and most peasants had less than one acre.

Production expanded rapidly after World War II. By 1954, there were 3,197 coffee growers producing 113 tons worth more than 35,000 pounds. In 1955, there were already 68 coffee nurseries able to provide sufficient seedlings to plant nearly 1,000 acres (KNA/DC/KSI/1/17). This was a time when the Gusii had also taken to the growing of pyrethrum, passion fruit and, of course, maize, which by 1950 had established itself as a major export crop. Nevertheless, the emergence of cooperative marketing, especially of coffee, attests to the importance of the crop in a maize-dominated area. The Kisii Coffee Growers' Cooperative was started in 1947 and grew to become a Union (Kisii Farmers Cooperative Union) in 1950, with primary societies based on the various pulping stations. In the same year, a lorry was acquired and stores built to enable the union to market member's coffee and other produce.

The first coffee factory in Gusiiland was built at Mogunga in 1952. By the mid-1950s, the bulk of the crop in South Nyanza district come from the Gusii highlands, with 26 out of 31 coffee societies being in Gusiiland (KNA/DC/KSI/1/22). The Gusii peasants took advantage of the removal of the maximum acreage limitations on coffee after 1954, and total production rose from 282 tons on 2,165 acres, grown by 5,763 farmers, in 1956 (KNA/KSI/1/18) to 4,400 acres, grown by about 19,000 farmers and earning them over 300,000 pounds in 1961 (KNA/KSI/1/23). By 1963, the crop was being grown by 36,140 framers with corresponding increases in acreage and income.

Coffee Production and Its Implications for Women

The increasing adoption of coffee production among the Gusii not only deepened social and economic stratification but also led to the intensification of women's labour while marginalising them in terms of ownership. Initially chiefs and educated people were given approval by the agricultural field staff to grow coffee after meeting certain standards of training in culturing, pruning, nursery work, planting, bench terracing and disease control. However, it was only the rich peasants, those in white-collar employment or migrant workers who were able to benefit. Almost all were men, leaving women to provide the much-needed labour in planting, weeding, picking and drying the seeds. Being in some form of employment, men devoted their time to this, while the women were left behind to tend the subsistence farms, and now the coffee farms, which in any case belonged to the men. This asymmetrical division of labour, whereby men tended paid labour outside the home while women remained at home as housewives to take care of the households and do other unpaid chores, worsened the position of women among the Abagusii. Mies (1998:68) has noted that, elsewhere in the colonial economy, the process of proletarianisation of men was also accompanied by a process of 'housewifisation' of women. Thus, men in employment tended to turn their wives into housewives, where their role was reduced to a producers of invisible goods and services without tangible monetary value to them. Thus, the new capitalist class of coffee growers rose on the subjugation of women.

As coffee growing gradually became popular, it came to be regarded as the 'modern' sector as opposed to the subsistence or maize sector. The cash crop sector came to be seen as the men's domain, just as the subsistence sector was largely identified with women. It was the goal of all men to enter into the cash crop sector, even those in work elsewhere. The greater impetus for coffee production came after the Second World War, when a great deal of wealth poured in through family remittances of conscripted soldier's gratuities, the sale of livestock and agricultural produce. The desire for profitable investment of this capital led to unprecedented demand for coffee production after the war. Most of those who joined the industry experienced a number of problems, including poor cultural conditions, inadequate pruning and spraying and lack of mulching. Diseases, the plucking of small under-ripe and yellow cherries and poor drying procedures combined to lower the quality of coffee. Nevertheless, coffee farming became the single most popular and productive crop for all men who had the resources to grow it. Thus, the progress of the coffee farmers was based on the subordination and exploitation of their own women. The law of progress, according to Mies, is always a contradictory one: progress for some means retrogression for others, and development of productive forces for some means underdevelopment for others. The reason why there cannot be a unilinear progress is that the predatory, patriarchal mode of production constitutes a non-reciprocal relationship (Mies 1998:76).

One might expect that, in such exploitative capitalist and patriarchal relations, women would have descended into hopelessness and total dependence on men. This was, indeed, the for a few women who could not fight back to prevent the further degeneration of their position under colonialism. However, many other women devised varied strategies to counter the negative forces of colonialism and assert their resourcefulness in the agricultural sector. Women were not merely victims of these processes. To assume so is to further reduce women to merely responding to external stimuli. Women were able to perceive these changes correctly and respond positively with a view to alleviating their position. Thus, women were capable of charting their own destiny rather than being merely helpless victims of the situation.

They managed to achieve this through several approaches. First, they deliberately intensified their own labour. As they were forced to undertake duties of their absent husbands, women had no alternative but to work a little more and longer than before. Long working hours became one other characteristic of industrious women. Secondly, they used the working parties more than before. Such working parties as *ekebosano, egesangio* and *ekiamorogoba* were redefined, not just to pool labour for tasks requiring colossal labour, but attained an economic motive as well. The working parties went around soliciting cash jobs on rich farmers' holdings. Thus, rich people, often those who had adopted coffee production, acquired extra labour for expanding production, weeding and plucking coffee berries. This earned the marginalised women some money, which they used for domestic requirements and other forms of investment, such as school fees for their children or group ventures (e.g., buying posho mills). Thirdly they sought employment locally in the rich men's *shambas* as individuals. This meant working for their employer in the morning hours and working on their own landholding in the afternoon. The intensification of women's labour cannot be underestimated in such circumstances.

The fourth strategy employed by the women to cope with their continued marginalisation from the cash crop economy was to increase production of profitable crops within their reach. Such women established vegetable gardens and were often seen selling vegetables in market places on appointed market days. Others grew fruits and sugarcane, which they also sold for extra earnings. The subsistence notion of producing what was required within the household was revolutionalised and came to be seen as a ready source of money for those not in the cash crop sector. For women who had been consigned to this subsistence sector, only enhanced innovation ensured their survival. Lastly, women formed small-scale cooperatives or 'merry-go-rounds' to raise the required capital. Out of their meagre earnings, each contributed for one of their members in a rotational manner so as to raise a substantial amount to enable the member make some vital investments. Evidently therefore, women charted their destiny amid unfavorable patriarchal and colonial conditions. Even the production of lucrative crops such as coffee came to rely on women as domestic, hired and contract workers, initially for free, but later for a wage or fee. Yet, all along, women remained the household custodians of food. All the food requirements within their households were their prime objective even with such communal labour undertakings, household chores, and engagement in wage labour.

Conclusion

This chapter has attempted to critically analyse the changing role of women in agricultural production in the face of agricultural innovations and, in particular, the introduction of coffee and its technology among the Abagusii. It has been argued that the role of women in agricultural production was fairly disadvantaged in the pre-colonial period, given the male dominance over the factors and relations of production. Further, it has been shown that the totality of the oppressive and exploitative relationship between men and women in pre-colonial Gusiiland was amplified by technological innovations accompanying the introduction of coffee in the area. The lucrative cash crop became the preserve of male farmers, while female farmers were relegated to subsistence crop production. This benefited the male farmers most, thus entrenching them in 'modern' agriculture, often with female labour. With the advent of labour migrancy, mainly to the neighbouring tea estates in Kericho, female farmers were further burdened, as they had to take up roles formerly done by men. Changes in women's participation in agricultural production occasioned by the introduction of coffee in Gusiiland have been tackled, showing the disadvantaged relations of production that women found themselves in. The response of women in coping with their continued marginalisation proves that they were capable of determining their destiny in various ways, taking into cognisance that Gusii women do not constitute a homogenous group, nor were they equally and uniformly affected. Their efforts to resist and change their marginal position in coffee farming portray them as agents of their own destiny rather than as victims of male patriarchy and dominance.

References

Barnes, C., 1976, 'An Experiment with African Coffee Growing in Kenya. The Gusii, 1933-1950', unpublished PhD. thesis, Michigan State University.

Boserup, E., 1970, *The Conditions of Agrarian Growth: The Economics of Agrarian Changes Under Economic Pressure,* London: Allen and Unwin.

Garst, R. D., 1972, 'The Spatial Diffusion of Agricultural Innovations in Kisii District, Kenya', unpublished PhD. Thesis, Michigan State University, Michigan.

Goodman, D. and Redcliff, M., 1981, *From Peasants to Proletarians,* Oxford: Blackwell.

Goonatilake, S., 1984, *Aborted Discovery: Science and Creativity in the Third World,* London: Zed Books.

Emeagwali, G., 2003, *Colonialism and Africa's Technology,* (http://www.africahistory.net/colonial.htm), 30 May, 2003.

Grieco, M., 2002, *Gender and Agriculture in Africa: The Expert Neglect of Local Practice.* (ww.geocities.com/margaret_grieco/femalefa/genagr.html). 30 May 2003.

Harding, S., 1998, *Is Science Multicultural?: Post-colonialisms, Feminisms and Epistemologies,* Bloomington: Indiana University Press.

Johnson, J. and Kelb, M., 1985, *Women as Food Producers in Developing Countries,* Los Angeles: UCLA African Centre

Lerner, G., 1986, *The Creation of Patriarchy,* New York: Oxford University Press.

Mies, M., 1998, *Patriarchy and Accumulation on a World Scale: Women in the International Division of Labour,* London: Zed Books.

Ochieng', W. R., 1974, *A Pre-Colonial History of the Gusii of Western Kenya AD 1500-1914,* Nairobi: East African Literature Bureau.

Olson, P., 1994, 'Feminism and Science Reconsidered: Insights from the Margins', in J. Peterson and D. Brown, eds., 1994, *The Economic Status of Women under Capitalism: Institutional Economics and Feminist Theory,* Aldershot: Edward Elgar.

Omwoyo, S. M., 1992, 'The Colonial Transformation of Gusii Agriculture', unpublished M.A. Thesis, Kenyatta University.

Omwoyo, S. M., 2002, 'Impact of Colonialism on Labour Patterns Among African Communities in Kenya: The Case of the Abagusii of Western Kenya', in *Eastern Africa Journal of Humanities and Sciences,* Vol.2, no.1.

Stichter, S., 1982, *Migrant Labour in Kenya: Capitalism and African Response 1895-1975,* London: Longman.

Zeleza, P. T., 1986, 'The Current Agrarian Crisis in Africa: Its History and Future', in *Journal of Eastern African Research and Development,* Vol. 16.

13

Gender-Based Associations and Female Farmers Participation in Science and Technology Projects in Anambra State of Nigeria

Anthonia I. Achike

Introduction

The protocols observed in government institutions often lead to delays in decision-making and implementation of plans, no matter how well articulated. The impersonal approach to government duties and the 'top-down' system of policy formulation and implementation also contribute to the failure of even well-articulated government programmes. Anambra State in Nigeria is no exception. Thus, many community-based development projects have become survival/coping strategies for various communities in the state, leading to the birth of Community Development Associations (CDAs). CDAs are organisations of different individuals with varying academic backgrounds and experiences, but with the common objective of developing their town or community through selfless efforts and volunteer services. Occasionally, government projects are implemented through CDAs, and governments rely on CDAs for grassroot coverage of their programmes. Some CDA projects are concerned with science and technology issues in agriculture, such as skills acquisition, genetically modified plant and animal varieties, germ-plasma development and, agro-chemicals. However, with the increasing number of CDAs, a high turnover rate in CDA formation and dissolution has emerged, with CDAs having an average lifespan of only three years. This is a course for concern, considering their societal obligations and challenges.

This high turnover rate may not be unconnected with the informal nature and generally weak organisational structures of CDAs.

On the other hand, the male-female dichotomy in societal roles and obligations has led to the formation of Gender-Based Associations (GBAs) from CDAs. GBAs are single-sex associations, essentially CDAs formed along gender lines. They also constitute a principal feature of the normalisation processes that characterise the on-going social, political and economic changes in rural and urban societies in Anambra. Since GBAs are closely associated with informal strategies and practices, it is likely that informalisation has potentially critical repercussions for gender advocacy, social equity and policy change (Esman and Uphoff 1984).

Current debate on informalisation, gender networks and social change revolves principally around the potentials and constraints of informal GBAs in promoting participatory development in science and technology, influencing policy change and re-defining gender. Over the years, disillusionment with the failures of top-down development approaches, along with the aggravated crises of state, economy and society, has fuelled demand for the democratisation of development (OAU/ILO 1983; Oakley and Marsden 1984; UNECA 1990). Within this framework, GBAs are expected to perform critical roles in the political economy, including acting as intermediaries to transform social and economic relations, channels to articulate and voice grassroots interests, and platforms to promote power devolution (Bratton 1987; Rahmato 1991). Furthermore, GBAs are envisaged as agencies to advocate and defend the gender space and as instruments to engender political consciousness and nurture participatory and equal development in science and technology (Holmsquist 1980; Chambers 1983; Esman and Uphoff 1984; Cernea 1987; Rahmato 1991).

The logic underlying the linking of GBAs to the larger project of social change and development in science and technology is that, since GBAs apparently bear attributes of legitimacy, autonomy and grassroots patronage, they should be able to expand their influence to serve as significant forces driving the new technological advances in agriculture. This assumption is largely untested in empirical terms and thus remains shrouded in uncertainty. The need to clarify the situation based on empirical evidence is the rationale for this study.

Informalisation and Restructuring

My theoretical framework revolves around the fact that informalisation, gender networks and social change are interrelated concerns associated with social, political and economic restructuring processes, especially since the late 1980s. With the rapid expansion of social and economic informality in the wake of the restructuring phenomena, societies and cultures appear to be experiencing redefinitions and reconstructions of gender notions, leading to new models of social relationships and male-female dichotomies in science and technology. These new social paradigms and gender transformations are embodied in social change processes driven by a number of forces, among which are gender-based associative structures. While these structures may be motivated by social security, group preservation and the coping needs of the different genders, the wave of informalisation strategies and practices sweeping across society constitutes an additional means for gender-based associations to operationalise gender space and participation in science and technology.

Furthermore, the imperatives of popular participation and the competition for scientific and economic space in present-day societies and cultures invoke different coping responses among various gender-based structures. Hence, an important aspect of gendered science is the interaction between the growth of informality, gender-associative structures and the dynamic of social relationships. The innovative instruments and strategies adopted by gender-based associations vary temporally and spatially according to intervening constraints and incentives defined by the environment. The mechanics of these variations — and the resulting social products (in terms of participatory development in science and technology) — are real questions that demand interrogation.

With the above in mind, a research project was designed to look into these issues. The general objective was to investigate the potential of gender-based associations to enhance participation of female farmers in science and technology projects. The specific objectives were to:

- describe and analyse group formation on gender lines.

- highlight the constraints, incentives and coping strategies that have influenced the surviving GBAs.

- identify and describe science and technology projects in agriculture.

- analyse participation in, and benefits from, science and technology projects in agriculture on the basis of the gender of participants.

- compare the performance and impact of selected GBAs in participatory development in science and technology projects in agriculture.

Each of the three senatorial zones of Anambra state formed a sampling zone. Using a multi-stage purposive selection technique, five community development associations, including three GBAs, which have functioned for a minimum of five years were selected for the study:

- Women Ministries of the Anglican Diocese on the Niger (female members only)

- Akwa Multipurpose Cooperative Farmers' Association (male and female members)

- Ancient 'Otu-Odu' (Ivory) Society (females only)

- Ancient 'Agbalanze' Society Onitsha (men only)

- Anambra Self-Help Organisation (females only).

In a broad sense, these associations function as non-governmental extension agents for government and research institutes.

Focus group discussions and interviews with members and leaders of each of the selected associations were the main data collection instruments. Two focus group discussions were held with each of the associations, and their leaders were separately interviewed. One focus group discussion each was held with participants and trainees in the associations' projects; this was in addition to a detailed inspection of facilities in the associations' workshops, centres or green/breeding houses. Since the data collected were essentially qualitative in nature, they were simply described. The responses regarding perceptions of, and attitudes towards, gender and community development associations in Anambra state can be summarised as follows.

Group Formation on Gender and Non-Gender Lines

The focus group discussions revealed that many groups were formed based on the sex of individual members. This gave rise to a proliferation of single-sex groups in the various communities. Group formation on gender lines can be traced to the ancient traditions of the Igbos (the major ethnic group in southeastern Nigeria) in which specific gender roles are delineated, and where the culture of male superiority and dominance is inculcated in the socialisation processes. Hence, women and girls are regarded as weak, inferior and depen- dent creatures. This distinctive gender stereotype is naturally transferred to group formations. Thus, female groups were formed to play supportive roles and implement decisions taken by men. Even in mixed-sex associations, within-group gender activities are organised. For instance, in many churches, even though membership is non-gender based, activities are carried out under separate men's, women's, boys' and girls' associations. The same ideology is portrayed in community-based associations such as age grades and community development unions. In contemporary society, where women have proved their mettle, this gender stereotype has generated a lot of conflict and clamour for gender space. Many gender-based associations now target the uplifting of women and enhancement of their participation in modern scientific techniques. In agriculture, for example, traditional food crops are classified as female crops, while cash crops are regarded as male crops. And because of the higher cash returns involved, men dominate mechanised and scientific operations in agriculture. This is typified by one of the mixed-gender associations studied, which had a large plantain/banana plantation for both the male and female members. However, with the introduction of the hybrid plantain/banana strain, a relatively a capital-intensive project, it was observed that the association guaranteed the male participants soft loans to adopt the new technology while their female counterparts had only indirect access to the loans. Female farmers had to be guaranteed through their husbands, sons or brothers. This development led to the formation of a female farmers' association, which guaranteed loans for its own members. The women farmers' association also helped members to procure fertilisers, insecticides and pesticides at subsidised rates. However, the capital base of the female farmers' association was relatively small, and this limited its ability to support the members in large-scale farming. The same limitation invariably weakened the potential of female GBAs.

Group formations along gender lines were found to predominate in the study area, mostly based on inherited traditions, but occasionally formed to fight gender discrimination. Where mixed-sex groups exist, activities are ultimately carried out along gender lines. This has perpetuated the practice of encouraging girls and women to adopt software technologies, while their male counterparts are socialised into hardware technologies.

Agricultural Science and Technology Projects in Gender-Based Associations

The major science and technology projects in agriculture being handled or managed by the selected GBAs are agro-based skill acquisition centres, information communication centres, improved/hybrid crop plantation projects and agro-service centres. These projects are generally geared towards the enhancement of livelihood and poverty eradication.

All the CDAs except Akwa Multipurpose Farmers' Association have skill acquisition centres that generally train both sexes in the different skills available. However, it was observed that the female participants are trained mainly in cloth weaving, soap/pomade making and word processing, while their male counterparts are trained in hardware technologies such as welding, machine fabrication and engine repairs. This corroborated the findings of Kitetu (1998) regarding physics classroom discourse practices and the construction of gendered identities in a Kenyan secondary school. The officials of the GBAs regarded the dichotomy in areas of specialisation as natural, and there was no effort to encourage the girls to learn 'hard' technologies. Indeed, the few girls that expressed an interest were indirectly discouraged.

The training lasted between two and three years, and graduates were awarded diploma certificates in their areas of specialisation. The GBAs then assist graduates by guaranteeing them for soft loans to set up their own businesses and supervising their projects until 75 percent of the loan is repaid. This buttresses the use of the social capital of an organisation in community development. For example the Anambra State Self Help Organisation gave out the sum of N1.5m annually as micro-credit to its clientele. Some of the trainers in the skill acquisition centres are foreigners or expatriates who are specialists in different skills. The trainers provide the technical skills, while the institutions supply the personnel and the environment. The quality of the training personnel and the managerial ability of the GBA operators enhance the potentials of these GBAs to successfully execute their projects.

An example of a crop plantation project carried out by a GBA was furnished by the Akwa Multi-purpose Cooperative Farmers Association This was a large extension outfit established by with technical and funding assistance from the International Institute for Tropical Agriculture (IITA). The linkage to international institutions increases the potential ability of CBAs to assist their clients. The major aim of this project was to extend improved skills in plantation agriculture. Through this project, hybrid plantain, cassava, maize and banana strains resistant to the devastating black sigatoka and mosaic diseases are disseminated to participating male and female farmers. The farmers were thus encouraged and taught the breeding methods and the post-harvest processes. The products are processed into wines, chips, flour, pastries, etc. On completion of the training, the certified participants are given guaranteed soft loans from government and some international agencies, like USAID, to establish their own farms. Some of the observed gender issues in the plantation project were as follows:

- There were more male participants in the breeding and marketing sections of the project.

- The female participants specialised mostly in the post-harvest processing of the products into chips, flour and pastries.

- Alhough the loans were guaranteed for all the successful candidates, more males than females benefited. However, the few female participants that benefited kept to the terms of the loans and hence had lower default rates than their male counterparts.

Another common type of CBA/GBA project involves agro-service centres. These centres have equipment for sale and warehouses for compounding, packaging and storing agrochemicals for sale at subsidised rates to their members and other farmers. The female GBAs that established agro-service centres did so mainly to stock farm inputs, especially fertilizers, for sale to female farmers who normally have poor access to these inputs.

The selected GBAs successfully handled their chosen projects for an average of three years, and their strength was drawn from the commitment and dedication of the members and the management cadre. Some of the GBAs are owned and managed by the church, while retired top civil servants manage others.

Incentives and Constraints

The major constraints highlighted by the gender-based associations were:

- lack of funds to start new projects and complete existing ones. The female GBAs were worse hit by this problem because they found it relatively more difficult to obtain loans directly, since they lacked collateral. This affected the relative potential of female GBAs to carry out their scheduled responsibilities, and occasionally led to the collapse of some of them.

- volatility of both micro-economic and macro-economic variables in the economy: Inflation, interest rates, taxes and import/export regulations in Nigeria generally, and Anambra state in particular, fluctuate arbitrarily, and this interferes with the planning and execution of projects.

- corruption and changing value systems: Although all the associations are development-oriented, there seemed to be elements of selfishness and self-aggrandisement on the part of the some key functionaries. This led to corrupt practices that weakened the association. The phenomenon was traced to changing societal values in which the emphasis on self-enrichment overrides many group objectives. The problem was found to affect the male GBAs more than the female GBAs. More male GBAs had cases of sharp practices and embezzlement of funds in court. Even the few surviving ones were threatened by power tussles in different dimensions.

- lack of political will and government support: Although the government of Anambra state was involved in a gigantic poverty eradication programme through one of the GBAs studied, and was using another association as an agricultural extension outfit, it had not been honouring its part of the agreement for these projects. Women Ministries International had to get a loan to complete its bakery factory (a skill acquisition centre) after many years of trying to persuade the government to provide the promised funds. The same thing applied to the farmer's association that was being used to extend the propagation of hybrid plantain and banana products. They had to construct their own breeding houses, although these were supposed to have been built by government. In addition, government was supposed to assist in guaranteeing 75 percent of the loans given to the beneficiaries of these projects under the Agricultural Credit Guarantee fund, but this agreement has usually been 'honoured in breach'.

The ability to survive these and other constraints had been a major source of strength for these associations. Thus, despite the above constraints, many GBAs have stood the test of time, and their survival strategies hinge on a number of incentives from within and, occasionally, from outside the associations.

Some GBAs have often attracted external funding from international donor organisations such as USAID and the World Bank. For instance, the Akwa Multipurpose Farmers' Association and the Anambra State Self Help Organisation received technical and funding assistance from IITA to establish their plantain/banana plantation and extend the required technical skills to participating farmers. Also, wealthy and influential members of the communities have instituted various forms of scholarships and foundations for youth and women's development through some of these gender-based associations. For instance, the Ikem Ozobia Foundation, aimed at ameliorating the sufferings of the poor, is partly operated through the Ancient 'Out-Odu' Society of Onitsha, a gender-based association.

Generally, the projects executed through these gender-based associations lead to the construction of permanent buildings and workshops and the acquisition of modern equipment, which in the long run galvanises and strengthens the associations. Thus, as the president of the Women Ministries put it, 'with these gigantic projects, equipments and installations, we cannot afford to collapse; we have stood the test of time and the sky is our limit'. And according to the Chairperson of the Otu-Odu society:

Our strength is in God and our coping strategy is sincere and transparent community service; we inherited our association as a strong group from our parents, and we shall hand it over as a stronger group to our children. Even government is depending on us for the execution and implementation of their policies and that shows our commitment to sincere service; and trust from both the government and the people—we have come to stay.

Performance of Gender Based Associations in Science and Technology Projects

A comparative assessment of the male and female trainees in the science and technology projects organised by the selected GBAs showed that:

- more females (76 percent) than males (24 percent) enrolled in project activities. Some activities, e.g., metal fabrication, machine maintenance, vulcanising, cash crop production and mechanized post-harvest processes have 93 percent male representation, while weaving, sewing, soap/pomade making and word-processing have 94 percent female participation.

- the males, though fewer in number overall, scored higher marks than the females. Even in the soap/pomade class, which has mostly female participants, a boy scored the highest mark. The trainers attributed the better performance of boys to higher intelligence.

- more females than males completed their programmes. This was attributed to societal pressures and unguided ambition. The quest to make money fast led more males to look for greener pastures in other spheres of life or to establish their own business without completing the training programme.

- more females than males adopted and practised the science and technology innovations learned during the training programme. This is despite the observed higher performance of males relative to females in the training programme. The adoption categories used were 'early adopters', 'late adopters' and 'non-adopters'. Females dominated the first two categories, while males dominated the third. The trainers attributed this to unguided ambition on the part of the male graduates. According to them, the quest to make quick money made most male graduates introduce 'shortcuts' in the skills they acquired and hence get classified as non-adopters of the skills. This however does not mean that the boys do not practice the acquired skills in the long run; it was rather a measurement problem of the adoption index used, which counted as adopters only those that strictly followed the sequence of application of skill. It may also be a methodological problem of measurement, possibly linked to the trainers having to report on the successful outcome of the training to the funders, rather than a reality of using the acquired skills. If so, this represents a capacity gap on the part of the trainers and calls for capacity strengthening.

- More females than males kept to the terms of their loans. About 80 percent of the loans guaranteed for the female graduates have been repaid compared to 32 percent of the loans for males.

- the female GBAs used more science and technology innovations than their male counterparts. However, the male GBAs also had other community development projects such as electrification, road construction and supply of pipe borne water to their rural communities, which the female GBAs appeared to de-emphasise. Each GBA had its own areas of emphasis and worked towards a defined goal, but their ability to achieve their goals varied with their financial and technical capabilities and also their national and international linkages.

Summary and Conclusion

GBAs are community-based, non-governmental organisations or voluntary networks formed for the purpose of community development. They are generally committed to the development of their communities through self-help projects. The strengths of GBAs hinge on their inherent power of networking, as highlighted in various studies of the power of social networks, which note, among other things, that the attributes of individuals are less important than their relationships and ties with other actors within the network.

The present study shows that some GBAs in Anambra State in Nigeria have specific tasks, while others concentrate on perceived areas of need such as agricultural development. Also, many of the GBAs have weak organisational structures that reduce their networking effectiveness. Some are not even registered with the Corporate Affairs Commission of Nigeria. However, despite these weaknesses, they have grassroot support and patronage. The government recognises them and some government policies are implemented through them. For instance, part of the National Poverty Eradication Policy (NAPEP) is implemented through Women Ministries International.

Female GBAs should be particularly encouraged because of their relative better performance and their efforts to enhance the status of women, who are usually discriminated against in science and technology development issues. Finally, although GBAs are good conduits for the extension of scientific innovations to farmers and implementation of many community development projects, they need to be further strengthened for sustainable development.

References

Bratton, M., 1987, 'Drought, Food and Social Organisations of Small Farmers in Zimbabwe', in M. Glartz, ed. *Drought and Hunger in Africa,* Cambridge: Cambridge University Press.

Chambers, R., 1983, *Rural Development: Putting the Last First,* London: Longman.

Cernea, M., 1987, 'Farmer Organisations and Institution Building for Sustainable Development', in *Regional Development Dialogue,* Vol. 8, no. 2, pp. 1-19.

Esman, M. and Uphoff, N., 1984, *Local Organisations: Intermediaries in Rural Development,* Ithaca: Cornell University Press.

Holmsquist, F., 1980, 'Peasant Political Space in Independent Africa', *Canadian Journal of African Studies,* Vol. 14, no. 1, pp. 156-67.

Kitetu, C., 1998, 'An Examination of Physics Classroom Discourse Practices and the Construction of Gendered Identities in a Kenyan Secondary School,' unpublished PhD. thesis, Lancaster University.

Oakley, P. and Marsden, D., 1984, *Approaches to Participation in Rural Development*, Geneva: International Labor Office.

OAU/ILO, 1983, 'Democratisation of National Development in Africa', proceedings of a symposium in Dakar, March 15-18, 1984, Geneva: International Labor Office.

Rahmato, D., 1991, 'Peasant Organisations in Africa: Constraints and Potentials', Dakar: CODESRIA.

UNECA, 1990, *Africa Charter for Popular Participation in Development and Transformation*, Addis Ababa: United Nations Economic Commission for Africa.

www.ingramcontent.com/pod-product-compliance
Lightning Source LLC
Chambersburg PA
CBHW021818270326
41932CB00007B/241